Graphene-Based Nanomaterial Catalysis

Edited by

Dr. Manorama Singh

Department of Chemistry,
Guru Ghasidas Vishwavidyalaya,
Bilaspur, Chhattisgarh-495009
India

Dr. Vijai K. Rai

Department of Chemistry,
Guru Ghasidas Vishwavidyalaya,
Bilaspur, Chhattisgarh-495009
India

Dr. Ankita Rai

School of Physical sciences,
Jawaharlal Nehru University,
New Delhi-110067
India

Graphene-Based Nanomaterial Catalysis

Editors: Manorama Singh, Vijai K Rai & Ankita Rai

ISBN (Online): 978-981-5040-49-4

ISBN (Print): 978-981-5040-50-0

ISBN (Paperback): 978-981-5040-51-7

need for a court order if at any point you breach any terms of this License Agreement. In no event will any delay or failure by Bentham Science Publishers in enforcing your compliance with this License Agreement constitute a waiver of any of its rights.

3. You acknowledge that you have read this License Agreement, and agree to be bound by its terms and conditions. To the extent that any other terms and conditions presented on any website of Bentham Science Publishers conflict with, or are inconsistent with, the terms and conditions set out in this License Agreement, you acknowledge that the terms and conditions set out in this License Agreement shall prevail.

Bentham Science Publishers Pte. Ltd.
80 Robinson Road #02-00
Singapore 068898
Singapore
Email: subscriptions@benthamscience.net

BENTHAM SCIENCE

CONTENTS

PREFACE

Electrocatalysis is a special field of electrochemistry and has been received tremendous evolution in the nineties due to its applications in chemistry, industrial, and chemical engineering. Using emerging and advanced materials for electrocatalysis in different applications is a promising field in research and development. Amongst, *graphene* and *its derived nanomaterials* are excellent materials to be used as electrocatalysts. Graphene-based nanomaterials are a novel class of nanomaterials that include graphene and its derived materials (prepared *via* functionalization with inorganic and organic nanomaterials). Extensive studies on their synthesis and applications have been conducted so far in various fields of chemistry, physics, biology, engineering, and applied sciences. Especially, graphene-based nanomaterials have many potential catalytic applications in electrochemical sensing, oxygen and hydrogen evolution, fuel cells, organic transformations, *etc*. Since the last two decades, incessant efforts of researchers worldwide have resulted in designing and synthesizing more novel graphene-based nanomaterials and have developed more knowledge about their promising catalytic applications in different fields.

This book delivers knowledge of the recent trends in using graphene-based nanomaterials as an electro/-organic catalyst for different promising applications. **Chapter I to Chapter X** describes the general introduction, properties, and various advanced synthetic approaches for the construction of graphene-based nanomaterials and the recent trends of their electrocatalysis towards various applications, *e.g.*, electrochemical sensing of various analytes, in advanced catalytic performance as heterogeneous catalysts in reduction reactions, carbon-carbon bond formations, carbon-heteroatom bond formations, multicomponent reactions, carbo-catalytic cycloaddition, stereoselective, ring-opening, and ring-closing reactions, as electrocatalysts in oxygen and carbon dioxide reduction, water splitting and fuel cells.

Strategic applications of these graphene-based nanomaterials as efficient carbo electro/-catalyst employed in a variety of advanced applications have been provided in this book. An updated and comprehensive account of the research in this field will provide the readers an opportunity to explore new dimensions for designing useful new graphene-based nanomaterials as well as unexplored electro-/organic catalytic reactions important in both academia and industries. The topic chosen in this valuable book will be beneficial for a broad range of readers such as graduate, postgraduate, Ph.D. students, faculty members, research & development (R & D) personnel working in these areas.

Of course, this book does not include all achievements and aspects of graphene nanomaterials. Yet, we hope that readers will gain profound insights into this fascinating research area through this book and be attracted to keep an eye on this field in the future, and even be inspired to join the search for new applications of graphene-based nanomaterials.

Manorama Singh
Department of Chemistry
Guru Ghasidas Vishwavidyalaya
Bilaspur, Chhattisgarh
India

Vijai K. Rai
Department of Chemistry
Guru Ghasidas Vishwavidyalaya
Bilaspur, Chhattisgarh
India

Ankita Rai
School of Physical Sciences
Jawaharlal Nehru University
New Delhi-110067
India

List of Contributors

Amisha Soni	Department of Chemistry, Indian Institute of Technology, Banaras Hindu University, Varanasi, 221005, India
Adel Saadi	Laboratory of Natural Gas, Faculty of Chemistry, USTHB, Algiers, Algeria
Ajit Sharma	Department of Chemical Engineering and Physical Sciences, Lovely Professional University, Phagwara 144411, India
Amel Boudjemaa	Laboratory of Natural Gas, Faculty of Chemistry, USTHB, Algiers Centre de Recherche Scientifique et Technique en Analyses Physico-Chimiques, Bou-Ismail CP 42004, Tipaza, Algeria
Angeliki Brouzgou	Department of Energy Systems, School of Technology, University of Thessaly, Geapolis, Regional Road Trikala-Larisa, 41500 Larisa, Greece Department of Mechanical Engineering, University of Thessaly, Pedion Areos, 38334, Volos, Greece
Benjamín Valdez	Instituto de Ingeniería, Mexicali, Universidad Autónoma de Baja California, Baja California, México
Bilal Chikh	Laboratory of Natural Gas, Faculty of Chemistry, USTHB, Algiers
C R Ravikumar	Research Center, Department of Science, East-West Institute of Technology, VTU, Bengaluru 560091, India
Deepak Kumar	Department of Chemical Engineering and Physical Sciences, Lovely Professional University, Phagwara 144411, India
Diganta Sarma	Department. of Chemistry, Dibrugarh University, Dibrugarh-786004, Assam, India
Dipika Konwar	Department of Chemical Sciences, Tezpur University, Napaam, Tezpur, Assam, 784028, India
Gisela Montero	Instituto de Ingeniería, Mexicali, Universidad Autónoma de Baja California, Baja California, México
H. C. Ananda Murthy	Department of Chemistry, School of Applied Natural Science, Adama Science and Technology University, Adama, P.O. Box.1888, Ethiopia
Imane Ghiat	Laboratory of Natural Gas, Faculty of Chemistry, USTHB, Algiers
Jasmin Sultana	Department. of Chemistry, Dibrugarh University, Dibrugarh-786004, Assam, India
K B Tan	Department of Chemistry, Faculty of Science, Universiti Putra Malaysia, 43400 Serdang, Selangor, Malaysia
Kah-Yoong Chan	Centre for Advanced Devices and Systems, Faculty of Engineering, Multimedia University, Persiaran Multimedia, 63100, Cyberjaya, Selangor, Malaysia
Leena Khanna	University School of Basic and Applied Sciences, Guru Gobind Singh Indraprastha University, Sector 16-C, Dwarka, New Delhi-110078, India
Manisha Malviya	Department of Chemistry, Indian Institute of Technology, Banaras Hindu University, Varanasi, 221005, India

Mansi	University School of Basic and Applied Sciences, Guru Gobind Singh Indraprastha University, Sector 16-C, Dwarka, New Delhi-110078, India
Mario A. Curiel	Instituto de Ingeniería, Mexicali, Universidad Autónoma de Baja California, Baja California, México
Mary T. Beleño	Instituto de Ciencias Agrícolas, Universidad Autónoma de Baja California, Mexicali, Baja California, México
Nayuesh Sharma	Department of Chemistry, Indian Institute of Technology Ropar, Rupnagar 140001, India
Nigussie Alebachew	Department of Chemistry, School of Applied Natural Science, Adama Science and Technology University, Adama, P.O. Box.1888, Ethiopia
Pankaj Khanna	Department of Chemistry, Acharya Narendra Dev College, University of Delhi, Kalkaji, New Delhi-110019, India
Prantika Bhattacharjee	Department of Chemical Sciences, Tezpur University, Napaam, Tezpur, Assam, 784028, India
R Balachandran	School of Electrical Engineering and Computing, Adama Science and Technology University, P O Box 1888, Adama, Ethiopia
Ricardo Torres	Instituto de Ciencias Agrícolas, Universidad Autónoma de Baja California, Mexicali, Baja California, México
Sandeep Kumar	Department of Chemistry, Indian Institute of Technology Ropar, Rupnagar 140001, India
Somit K Singh	Department of Chemistry, College of Natural and Mathematical Sciences, The University of Dodoma, PO Box 259, Tanzania
Sarvatej Kumar Maurya	Department of Chemistry, Indian Institute of Technology, Banaras Hindu University, Varanasi, 221005, India
Utpal Bora	Department of Chemical Sciences, Tezpur University, Napaam, Tezpur, Assam, 784028, India

CHAPTER 1

Introduction of Graphene-based Materials (Structure, Synthesis, and Properties)

Mary T. Beleño[1], Gisela Montero[2], Benjamín Valdez[2], Mario A. Curiel[2] and **Ricardo Torres[1,*]**

[1] *Instituto de Ciencias Agrícolas Universidad Autónoma de Baja California, Mexicali, Baja California, México*

[2] *Instituto de Ingeniería, Universidad Autónoma de Baja California, Mexicali, Baja California, México*

Abstract: Graphene and its derivatives are being studied in almost all fields of science and engineering. In recent decades, graphene has emerged as an exotic material and has received considerable attention due to its exceptional physicochemical properties, electron mobility, mechanical resistance, high surface area, and thermal conductivity. Graphene has a flat monolayer of carbon atoms (2D structure). The carbon-carbon bonds have sp^2 hybridization and are arranged in a hexagonal crystal lattice in the form of a honeycomb. It is the building block of all other graphite elements, including graphite itself, carbon nanotubes (CNTs), and fullerenes. Herein, we present an overview of graphene's classification, structural characteristics, and its chemical, physical, and technological properties. The synthesis routes are also classified according to the graphene precursors. The vast majority of the methods (button-up and top-down) currently used to obtain graphene and its derivatives are described. In addition, we provide a brief overview of methods of functionalization of graphene. The functionalization of graphene can be performed by covalent and non-covalent modification techniques. In both cases, surface modification of graphene oxide followed by reduction is carried out to obtain functionalized graphene.

Keywords: Allotropes, Graphene, Overview of graphene, Synthesis of graphene.

INTRODUCTION

The term nanomaterial is relatively new. Nanomaterials, materials structured on a nanometric scale (of the order of 10^{-9} m), can be obtained from different elements or chemical compounds [1]. Carbon nanomaterials are classified as carbon nanostructures, like carbon nanotubes, diamond, graphene, *etc*. Among these

---

* **Corresponding author Ricardo Torres:** Instituto de Ciencias Agrícolas Universidad Autónoma de Baja California, Mexicali, Baja California, México; E-mail: ricardo.torres26@uabc.edu.mx

Manorama Singh, Vijai K. Rai and Ankita Rai (Eds)

carbonaceous materials, graphene has attracted great research interest due to its unique structural characteristics and excellent performance. Furthermore, the production cost of graphene is very low compared to other carbon-based nanomaterials. Graphene is one of the allotropes of elemental carbon (carbon nanotubes, fullerene, diamond, *etc*.). The extremely extensive allotropy of carbon is due to the carbon atoms' ability to form highly complex networks of various structures [2].

Graphene can be extracted from graphite and is simply a sheet of graphite. It is a flat monolayer of carbon atoms (structure 2D) with a carbon-carbon bond length of 0.142 nm. The sp^2 hybridized carbon bonds are arranged in a honeycomb-shaped hexagonal crystal lattice. The existence of this nanomaterial has been known for a long time. In 1962, in an effort to synthesize graphene, Boehm and colleagues separated thin sheets of carbon by heating and chemically reducing graphite oxide [2 - 5]. However, they did not generate perfect monolayer graphene. Until 2004, Andre K. Geim and Konstantin S. Novoselov from The University of Manchester used a method to isolate highly oriented pyrolytic graphite (HOPG) and monolayer graphene. The latter is achieved by isolating the first carbon atomic thickness flakes and repeatedly peeling pyrolytic graphite with adhesive tape until graphene is obtained. This approach provides high-quality graphene comprising hundreds of microns [6].

Andre Geim and Konstantin Novoselov won the 2010 Nobel Prize in Physics for their groundbreaking work on graphene. The growing interest in graphene is mainly due to several exceptional properties that this material possesses. Since then, graphene research, including the control of graphene layers on substrates, the functionalization of graphene, and the exploration of graphene applications, has grown exponentially, with a sharp increase since 2004 [7].

According to the dimension of the space that a material occupies, the classification can be made from zero-(0-), one-(1-), two-(2-) dimensional, to the bulk (3D). Graphene represents a conceptually new class of materials that are only one atom thick, the so-called two-dimensional (2D) materials (2D materials extend in only two dimensions: length and width; since the material is only one atom thick, the third dimension, height, is considered zero). Graphene is the basic 2D unit or building block of other graphic materials (Fig. **1**). If we wrap the layers of carbon atoms around them like the lining of a ball, arched in zero-dimensional structures (0D), we obtain fullerenes. If we roll them cylindrically into one-dimensional structures (1D), they will give rise to nanotubes; finally, if we superimpose more than 10 layers three-dimensionally (3D), we will obtain graphite [2, 8].

Graphene and its derivatives have unique physicochemical, mechanical, optical, electronic, and thermal properties. These compounds can be used in a wide range of applications, in various fields, including catalysis, photocatalysis, electronics, and biomedicine. Due to their high specific surface area, accessibility to the surface, and high adsorption capacity compared to any other material of this type, graphene exhibits potential applications in catalysis as a valuable substrate to interact with various species [10].

0D   **1D**   **2D**   **3D**

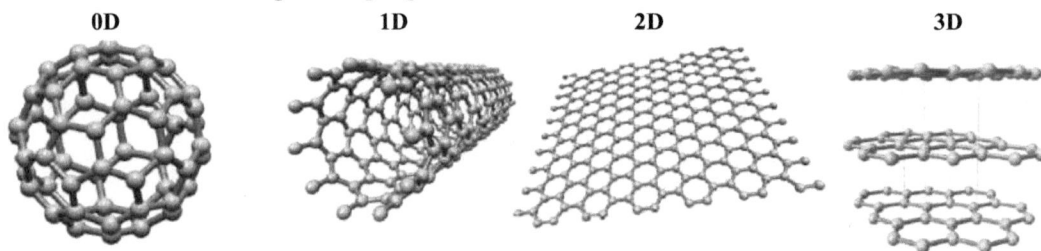

Fig. (1). Graphene: Basic component of other forms of carbon. Graphite is a stack of graphene layers (3D), graphene consists of a hexagonal lattice of carbon atoms (2D), CNTs are rolled-up cylinders of graphene (1D), and a buckminsterfullerene (C60) molecule consists of graphene balled into a sphere by introducing some pentagons as well as hexagons into the lattice (0D) [9].

Graphene-based composites are currently a focus of research, as their structural fabrications present better properties that would lead to newer applications. Recently, for example, graphene has been used as an alternative carbon-based nanofiller in the preparation of polymeric nanocomposites [11]. Adding graphene as a reinforcing agent in a polymer matrix has improved the overall performance and properties of such composites [11 - 13]. On the other hand, current research shows that graphene is capable of replacing metallic conductors in electronic and electrical devices due to its excellent electrical conductivity, mechanical flexibility, and high optical transparency. These properties allow graphene to be applied to various optoelectronic devices, from solar cells to touch screens [14, 15]. Graphene also has applications as a protective coating due to its unique shape and characteristics such as superconductivity, light-weight, high rigidity, and anti-corrosion ability [16].

Graphene is also studied as a material for biomedical use, in part, due to its high Young's modulus, high resistance to fracture, high electrical conductivity, and excellent optical performance. In addition, due to its large specific surface area, it is valuable for the adsorption of bioactive compounds. It has high optical absorption efficiency in the near-infrared (NIR) region compared to other types of carriers [17,18]. Various specific review articles are available on the properties of graphene in detail [19 - 23].

The applications mentioned above are made by modifying graphene based on defect modulation, during which specific types of disorders are quantitatively created to alter the crystal structure of graphene and consequently obtain the desired or improved properties.

These properties of graphene make it a versatile material, offering a wide range of possibilities for its use and commercialization. In 2013, the size of the graphene market was estimated at around the US $ 12 million [24]. Based on a (compound annual) growth of 51.7% between 2017 and 2022, the graphene market will reach the US $ 986.7 million in 2022 [25]. This book chapter presents a general and recent description of graphene and its derivatives, as well as its structural characteristics, outstanding properties, and routes for synthesis.

GRAPHENE STRUCTURE

Graphene has a nanometric structure in which carbon atoms are arranged in a hexagonal pattern to form a very compact sheet-like planar (2D) structure. Carbon atoms make bonds with surrounding carbon atoms with sp^2 hybridization, forming a benzene ring in which each atom donates an unpaired electron. The distance between neighboring carbon atoms in graphene has been reported to be approximately 1.42 nm [26]. There are three strong σ bonds (stronger than diamond ones) in each lattice that function as the rigid backbone of the hexagonal structure. The out-of-plane π bonds provide a weak Van der Waals interaction between the different adjacent graphene layers in 2LG and FLG [27].

Graphene stability is due to its tightly packed carbon atoms and sp^2 orbital hybridization—a combination of orbitals s, px, and py that constitutes the σ-bond. The final pz electron makes up the π-bond. The π-bonds hybridize together to form the π-band and π*-bands, which are bands crossing at the Fermi level. These bands are responsible for most graphene's notable electronic properties *via* the half-filled band that permits free-moving electrons [26]. This property is responsible for the higher electrical conductivity of graphene derivatives. For this reason, graphene is considered a zero band gap semiconductor, which has a small overlap between the valence and conduction bands [28]. The electrical conductivity of films can be modulated by applying an electric field, and this field effect supports the mechanism of most semiconductor materials. This effect is not possible in three-dimensional metallic structures because of the large number of free electrons present in metals that protect the electric field at an atomic distance [29].

Graphene is one atom thick (monolayer), which means that it is extremely thin. Various researchers have measured the thickness of graphene from 0.35 nm to 1.00 nm [30, 31]. Novoselov *et al.* have determined platelet thicknesses of 1.00

−1.60 nm [6]. Gupta *et al.* have measured the thickness of single-layer graphene film by atomic force microscope (AFM) as 0.33 nm [32]. This means that graphene is 1/200,000 the diameter of human hair.

The final shape of the graphene product depends, first of all, on the number of carbon layers that make up the material. Although "pristine" graphene is only one atomic layer thick, it is classified differently in the literature. The most common classes are as follows [33, 34]: very-few-layer graphene (vFLG, 1-3 layers), few-layer graphene (FLG, 2-5 layers), multi-layer graphene (MLG, 2 - 10 layers), or graphene nanoplatelet, which are stacks of graphene sheets that can consist of several layers (GNP, >10 layers).

According to the International Organization for Standardization (ISO) (ISO / TS 80004-13: 2017), the chosen terminologies to explain graphene and its derivatives are based on (ISO/TS 80004-13:2017) and are described as follows:

- Graphene: a single layer of carbon atoms (also called monolayer graphene or single-layer graphene); abbreviated as 1LG or SLG.
- Bilayer graphene: two well-defined stacked graphene layers; abbreviated as 2LG.
- Low-layer graphene: 3 to 10 well-defined stacked graphene layers; abbreviated as FLG.
- Graphene nanoplatelet or graphene nanoplates: graphene with lateral dimensions ranging from ∼ 100 nm to 100 μm and thickness between 1 and 3 nm; abbreviated as GNP [35].

In addition to carbon layers based on lateral size and structural characteristics, graphene-based materials can be further divided into graphene quantum dots (GQD), nanographene, graphene nanoribbons (arm–chair/zig–zag), SLG, MLG, oxide graphene (GO), graphene oxide quantum dots (GOQDs), reduced graphene oxide (rGO), and functionalized graphene. These add chemical species to the surface or edges of graphene for various applications [35, 36].

SYNTHESIS OF GRAPHENE

Currently, there are a large number of routes for the synthesis of graphene and graphene materials. These routes have been grouped into two categories according to the reagents or precursors used in the synthesis. The categories are as follows: top-down production and bottom-up production. In bottom-up production, the precursors are carbon gases, aromatic hydrocarbons, polymers, among others. While in top-down production, the synthesis starts with graphite [35, 37].

Top-down graphene production is relatively easy to perform, requires no substrate transfer, and is highly reproducible. This production method is less expensive and more profitable than bottom-up production, which requires sophisticated infrastructure and operating conditions. On the other hand, unlike top-down, bottom-up methods allow the formation of graphene to be controlled more precisely. Bottom-up production offers greater control of synthesis parameters, including the shape and size of graphene, resulting in high-quality graphene [35].

Each of the aforementioned categories presents advantages and disadvantages. However, the large-scale production of graphene and graphene-based materials present higher feasibility in top-down synthesis routes. Each of the categories and their different synthesis routes is described in detail below.

Bottom-Up Production

The synthesis of graphene or graphene-based materials using the bottom-up approach involves using hydrocarbon compounds as precursors [38]. In this approach, the precursors are usually in the gaseous state. Some of these are CH_4, C_2H_2, C_2H_4, C_2H_6, and C_3H_8. More recent investigations have taken carbonaceous materials in a solid and liquid state as precursors [35, 37]. The most important synthesis methods under the bottom-up approach are shown in Fig. (**2**) and described in detail below.

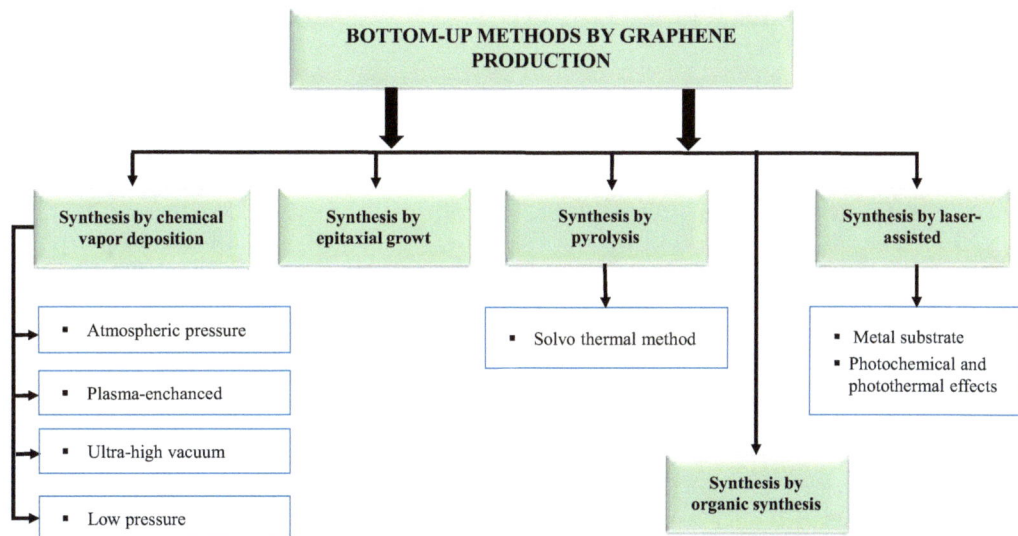

Fig. (2). Graphene synthesis methods under the bottom-up approach.

Synthesis by Chemical Vapor Deposition (CVD)

Chemical vapor deposition is one of the most popular methods for synthesizing carbonaceous nanomaterials because it is relatively simple and offers reasonably good control [38, 39]. The CVD technique produces different graphene products, from SLG and FLG to GNPs, using low-cost synthetic or natural graphite as a starting material [35]. This method involves the deposition of gaseous reagents on a substrate [40] inside a reaction chamber with controlled conditions of pressure and temperature. When gaseous reactants combine under the right conditions, a film of material (graphene) forms outside the substrate. The temperature of the substrate is the determining factor, which can vary depending on the type of substrate. The substrates used can be metals [41], alloys, oxides, among others [42]. CVD can be carried out at low pressure, ultra-high vacuum, plasma-enhanced or atmospheric pressure. A disadvantage of vapor deposition is that very little coating is done on the substrate and at a very low speed, often described as in microns of thickness per hour [35].

Synthesis by Epitaxial Growth

Among the various methods of growing graphene on surfaces, epitaxial growth is one of them. The preparation of graphene can occur by applying heat and cooling a silicon carbide (SiC) crystal. In general, on the Si face of the crystal, there will be single or two-layer graphene. However, on the C face, few graphene layers are grown [43]. This method is considerably sensitive to temperature and pressure and easily leads to the generation of CNTs instead of graphene if it experiences too high pressure and temperature [38, 44].

Synthesis by Pyrolysis

Graphene pyrolysis offers a cheaper and easier approach compared to epitaxial growth. This process involves the pyrolysis of a carbon precursor, such as a polymer, an oligomer, or a prepolymer, under solvothermal conditions to form graphene [45]. While this technique can produce high-purity functionalized graphene at low temperatures, significant defects in graphene properties have also been reported [46].

Synthesis by Organic Synthesis

The chemical synthesis of graphene is based on the assembly of basic atomic or molecular components, such as aromatic molecules from the oxidative cyclodehydrogenation of polyphenylene and oligophenylene sources. This route of synthesis produces small molecules of graphene and nanographene. Atomically precise and uniform nanostructures are obtained; it is suitable for the manufacture of precise and reproducible components for optoelectronic, nanoelectronic, and spintronic applications. This method has a disadvantage which is related to high production costs [35].

Laser-assisted Synthesis

Laser-assisted graphene synthesis is one of the fast-growing graphene production methods for specialized electronic applications. It can produce specific shapes in one step using two different approaches. The first approach is similar to the process, where a carbon precursor is first dissolved in a metal substrate by laser irradiation and, after cooling, precipitates as graphene on the surface of the metal substrate. This method uses nickel as a substrate and produces bilayer or multilayer graphene [47]. The second method produces porous graphene through photochemical and photothermic effects created by laser irradiation. This approach is performed in the absence of a metallic substrate [35, 48].

Top-Down Production

The synthesis of graphene or graphene-based materials can be carried out using the top-down approach. This strategy is defined as the strategy which is depending on the powdered raw graphite attack. The attack will eventually separate its layer to generate graphene sheets (from SLG and FLG to GNPs). This approach involves reduction, exfoliation, and oxidation processes, and its starting material is graphite [35]. The most important synthetic routes using this approach are presented in Fig. (**3**) and detailed below.

Synthesis by Chemical Oxidation-Reduction

The synthesis of graphene or graphene-based materials using the chemical oxidation-reduction method is a two-stage process. The first stage corresponds to the oxidation of the graphite followed by the sonication of the graphite oxide in a solvent such as water or ethanol to obtain GO. The second stage corresponds to the reduction of GO to obtain graphene or rGO. rGO is a form of GO that has been reduced to minimize the oxygen content and repair and restore defects in the GO.

The first stage is oxidative chemical exfoliation, which is considered one of the most promising routes for preparing graphite oxide and its derivative GO. The importance of this stage lies in the fact that GO is the raw material commonly used for the large-scale production of graphene materials. The synthesis of GO has been carried out conventionally following five methods, which are named after the researchers who proposed them: Brodie, Staudenmaier, Hofmann, Hummers, and Tour [35]. These methods have been further modified and improved to pursue an environmentally friendly and straightforward route. These methods involve the chemical reaction between graphite and a strong concentrated acid, such as sulfuric, phosphoric, and nitric acid. Then, the oxidation of basal and edge carbons by adding oxidant compounds such as potassium chlorate, potassium permanganate, and sodium nitrate is performed. The difference between the different proposed methods lies in the type of chemical compounds and the proportions of reagents used. During oxidation, oxygen-containing groups are introduced and spread throughout the carbon structure. This is done to overcome the van der Waals forces that hold the graphite sheets together and increase the space between the layers. A disadvantage of this synthesis route lies in the long reaction time and a strong emission of toxic gases during the reactions involved [38].

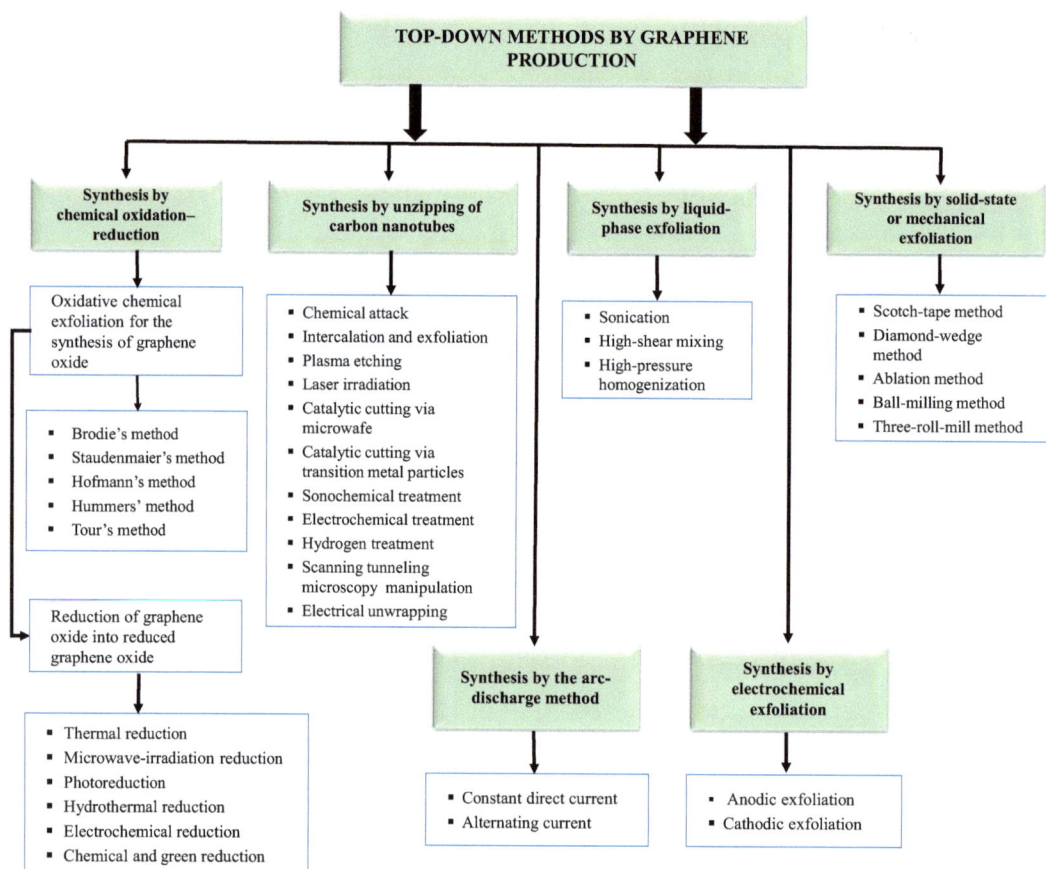

Fig. (3). Graphene synthesis methods under the Top-Down approach.

Synthesis by Unzipping of Carbon Nanotubes

An important route for the synthesis of graphene consists of the decompression of CNTs. Cutting the CNTs produces high-quality sheets of graphene. This synthesis route has a substantial economic advantage, consisting of low production costs, easy manufacturing, and high feasibility of large-scale production [49]. Unzipping is made possible by various processes, including chemical attack, intercalation, exfoliation, plasma *etc*hing, laser irradiation, catalytic cutting using microwaves or transition metal particles, sonochemical treatment, electrochemical and hydrogen treatment, *in-situ* Scanning Tunneling Microscopy (STM) manipulation, and electrical unwrapping synthesis methods. The graphene obtained is suitable for integration applications of electronic or optoelectronic devices [35, 49].

Synthesis by Arc-discharge Method

The synthesis of graphene using the arc discharge process is one of the most interesting methods because it can produce high-quality pure graphene and graphene doped with heteroatoms (N, B, and F) [50]. This method uses an electric arc furnace consisting of a graphite rod anode and a cathode. These are introduced into a water-cooled, steel-built vacuum chamber. There can be different gases in the chamber's reaction atmosphere; for example, hydrogen, helium, carbon dioxide, argon, and others. These gases are used to evaporate the discharge from the graphite bar and directly influence the type of graphene obtained (pure or doped) [50].

The arc discharge method can be carried out using constant direct current (DC) or alternating current (AC). This current flows through the electrodes when they make contact, resulting in the formation of plasma as well as vaporization of the anode. Graphene forms in plasma and deposits on the side of the chamber as black soot. There are several main drawbacks of the arc discharge method. Some examples are lack of control over graphene formation, unwanted doping, low amounts of elemental doping, and difficulties forming large electrodes [50].

Synthesis by Liquid-phase Exfoliation (LPE)

This synthesis route includes a wide range of liquids that can be used for the exfoliation of natural graphite and synthetic graphite in bulk to obtain SLG or FLG. Water, surfactant/water solutions, ionic liquids, organic solvents, and aromatic solvents are liquids used as exfoliation media for the exfoliation of bulk graphite [51]. In 2008, Coleman *et al.* proposed this synthesis route using sonication for the first time. At present, other liquid phase exfoliation methods are known, among which are: microfluidization, jet cavitation, high-shear mixing, and high-pressure homogenization. Each of these routes has its advantages and disadvantages. Some common advantages are that they do not use a cumbersome chemical oxidation step, are easy to operate, and do not require sophisticated instruments or high temperature or vacuum conditions. Furthermore, synthesis by liquid-phase exfoliation is considered a viable option to realize affordable graphene mass production [51, 52].

Synthesis by Solid-state or Mechanical Exfoliation

During solid-state exfoliation, the graphene layers are peeled from the graphite surface by mechanical means. Normal forces and lateral forces are used as an exfoliation mechanism to overcome the van der Waals forces between the graphite layers. A solid-state exfoliation can be performed in various ways. The first known synthesis route was the adhesive tape method proposed by Gein *et al.* [6, 53], which received the Nobel Prize in Physics in 2010. Other known routes are the diamond wedge method, ablation method, ball-milling method, and three-roll-mill method. Mechanical exfoliation methods are cost-effective and quickly and easily create high-quality graphene monolayers without the use of sophisticated equipment. However, it is labor-intensive, and the production throughput is low. Therefore, its use has been limited to laboratory-scale production [35, 38].

Synthesis of Graphene by Electrochemical Exfoliation

The electrochemical exfoliation method is one of the most environmentally friendly known methods and requires less processing time. This method generally involves the use of a solid or aqueous electrolyte and electrical current to provide structural exfoliation or expansion of graphite/graphene at a graphite anode or cathode. Different forms of graphite (graphite powder, rods, flakes, foils, or plates) can be employed as working electrodes in electrolytes during electrochemical exfoliation [54]. The drawbacks with this synthesis route generally occur when high exfoliation anode potentials are used around 510 V. High potentials cause intercalation of electrolyte ions and electrochemical reactions to occur rapidly, causing defects in graphene. These problems can be overcome using exfoliation with cathodic potential. However, exfoliation with cathodic potential has relatively low efficiency for the production of monolayer and bilayer graphene [54].

CHEMICAL AND PHYSICAL PROPERTIES OF GRAPHENE

The growing interest in graphene research is due to the fascinating properties exhibited by graphene. These make it a unique material with enormous potential for application in industry and technological devices. The excellent mechanical, electrical, and electronic properties of graphene derive from its unique 2D crystalline structure. They also depend on the number of graphene layers in graphene sheets. The horizontal dimension of graphene can be extended enough, while the thickness is only on the atomic scale (monolayer). Therefore, the structural characterizations of graphene must consider the horizontal macroscopic scale as well as the analysis at the atomic level [55, 56].

Graphene is a material that exceeds the hardness of diamond; it is considered the strongest material ever tested. It is at least 300 times stronger and more resistant than A36 structural steel of the same thickness. The single-layer graphene (SLG), free of defects, has a resistance of 42 N m^{-1} (breaking strength), which is equivalent to an intrinsic tensile strength of 130.5 GPa [57]. Breaking strength is defined as the maximum force that material free of defects can withstand before failing [58]. According to Kudin *et al.*, Young's modulus and Poisson's ratio for graphene are 1.02 TPa and 0.149, respectively [59, 60]. This indicates that the thickness of the graphene sheets has an important effect on the properties. The reported physical and chemical properties of graphene are summarized in Table **1**.

Table 1. General description of the physical and chemical properties of graphene.

Physical and Chemical Properties	Approximated Values	Ref.
Band gap	~ Zero	[28]
Covalent bonding energy	~ 5.9eV	[55]
Melting point	~ 4510K	[55]
Current density	~ 1.6 x 10^{-9} A cm^{-2}	[55]
Energy gap	~ 0.26eV	[55]
Shear modulus	~ 0.213 − 0.233	[55]
Tensile strength	~ 1100 GPa	[57]
breaking strength	~ 42 N m^{-1}	[57]
Young's modulus	~ 1.02 TPa	[59]
Poisson's ratio	~ 0.416 − 0.419	[60]
Specific surface area	~ 2630 m^2 g^{-1}	[61]
Optical transparency	~ 97.7%	[62]
Sigma bonds with a bond length	~ 1.42 nm	[63]
Charge carrier mobility at electron density ~ 2x10^{11}cm^{-2}	~15,000 − 200,000 cm^2 V s^{-1}	[63]
Electrical conductivity	~ 7200 S m^{-1}	[63]
Thermal conductivity	~ 5000W mK^{-1}	[63]
Fracture toughness	~ 4 − 5 MPa	[66]

On the other hand, graphene has a high specific surface area (2630 m^2 g^{-1}), in contrast to CNTs (1315 m^2 g^{-1}). With its high theoretical surface area, graphene provides a rich platform for surface chemistry and makes it an attractive candidate in energy storage applications [61, 62]. A frequently cited property of graphene is its high electron transport capacity. The unique atomic organization of carbon atoms in graphene allows its electrons to easily travel at a considerably high speed without the significant possibility of scattering, saving valuable energy normally lost in other conductors. The electrical conductivity of graphene at room temperature is 7200 S m^{-1} with average electronic mobilities that can reach ~ 200,000 cm^2 Vs^{-1} [63].

The thermal conductivity of single-layer graphene at room temperature is 5000 W mK^{-1} (being several times that of copper or silver), allowing it to disperse heat easily and withstand strong electrical currents without heating [63]. In terms of optical properties, graphene is a highly transparent material. Sun *et al.* reported that a single layer exhibits an optical transmittance of 97.7% recorded at a wavelength of 550 nm. In other words, it can absorb only 2.3% of incident visible light. Light absorption can increase as the number of overlapping graphene sheets increases [62]. The transmittance (λ = 550 nm) of bilayer graphene is 94.3%, and six-layer thick graphene has a transmittance of 83% at a similar wavelength [64, 65].

Due to the two-dimensional nature of graphene, the chemistry and reactivity of SLG are dominated by its surface and thickness. On the other hand, the GNP is dominated by its edges. Spectroscopic tests showed that the reactivity of the edges is at least two times higher than the reactivity of the individual graphene sheet in bulk [67]. According to Loh *et al.*, The chemical reactivity of graphenes lattices' geometrically strained regions is much higher than other regions [68]. This is attributed to the easier displacement of electron density above the plane of the ring. The zig-zag edges of graphene have also been found to show higher chemical reactivity compared to arm-chairs edges. As the formation of aromatic sextets is disturbed in the zig-zag edges, it leads to thermodynamic instability [68].

As mentioned above, graphene is aromatic in nature; therefore, it has a very dense electron cloud above and below the plane. Due to this, the molecular orbitals of organic molecules can easily interact with the π electrons of graphene. Consequently, the electrophilic substitution of graphene is much easier than nucleophilic substitution. According to Loh *et al.*, graphene can take part in certain classes of reactions, including cyclo-additions, click reactions, and carbene insertion reactions [68]. However, reactions on graphene's surfaces hamper its planar structure. Destruction of the sp^2 structure leads to the formation of defects and loss of electrical conductivity.

FUNCTIONALIZATION OF GRAPHENE

The functionalization and dispersion of graphene sheets are of crucial importance in their applications. Pure graphene sheets are mostly unreactive, but the surface functionalization makes them reactive with other materials [56]. Research on the functionalization of graphene is increasing, evidenced by several publications that appeared in the last decade. However, there are still two shortcomings that researchers are struggling to overcome for actual graphene applications. On the one hand, because of the zero band-gap energy, graphene transistors have a small on-off ratio [69]. On the other, exfoliated graphene cannot be dissolved or dispersed in both aqueous and organic solvents [70]. Functionalization, also known as doping, can be defined as the route by which new properties, purposes, structures, or abilities are added to a substance through the alteration of the material in the aspect of the surface chemistry. For example, highly conductive functional groups can be incorporated to advance the electrical conductivity of the targeted graphene structure [9]. Functionalization can be carried out by attaching particles or nanoparticles to the surface of a substance. This can be either by chemical bonding or by adsorption [56].

The functionalization of graphene or GO can be performed by non-covalent and covalent alteration techniques to achieve a stable dispersion in its dispersive medium. Both techniques share a similar process, which involves the superficial alteration of GO followed by reduction to rGO [56]. Since pristine graphene is chemically inert, GO, or rGO is commonly used as a functionalization precursor due to the presence of multiple oxygen-containing functional groups such as hydroxyl (-OH) and carboxyl (-COOH) on its surfaces. Furthermore, the exceptionally large flat surface of GO may provide the reaction sites necessary to anchor various chemical species during functionalization, where alteration of structure can take place at the surface or edges of GO [71].

The covalent functionalization method is based on altering the chemical structure by reaction between the oxygenated groups in GO or rGO and the functional groups of the molecules that want to be grafted onto the surfaces [71, 72]. This modification can take place at the end of the sheets or on the surface. Surface functionalization is associated with the rehybridization of one or more sp^2 carbon atoms of the carbon lattice in the sp^3 configuration accompanied by the simultaneous loss of electron conjugation [56]. GO is modified by chemical functionalization with amines, esters, isocyanate, and polymer shells or by electrochemical modification with ionic liquids. There are three forms of covalent functionalization: condensation reaction, substitution reaction, and addition reaction. Table **2** shows different types of graphene/GO covalent modification using different modifying agents in various dispersing media. According to recent reports, nucleophilic substitution is the best route for large-scale production of functionalized graphene from GO in the covalent functionalization method. It can be easily carried out in an aqueous medium and at room temperature. However, the electrical conductivity of functionalized graphene is significantly reduced with this type of modification [64].

Table 2. Different ways for functionalization of GO using different modifying agents [71].

Modification Methods	Modifying Agent	Dispersing Medium	Ref.
Condensation reaction	Adenine, cysteine, nicotinamide	Water	[73]
	Poly (vinyl alcohol) (PVA)	Water, dimethyl sulfoxide (DMSO)	[72]
	Triphenylamine-based poly(azomethine) (TPAPAM)	Tetrahydrofuran (THF)	[74]
	Octadecylamine (ODA)	THF, tetrachloromethane (CCl4), 1,2-dichloroethane	[75]
	Organic isocyanate	Dimethyl-formamide (DMF), N-Methyl-2-pyrrolidone (NMP), DMSO, hexamethylphosphoramide (HMPA)	[76]
Substitution reaction	4-Bromo aniline	DMF	[77]
	Polyglycerol	Water	[78]
	8-Anilinonaphthalene-1-sulfonic acid (ANS)	Water	[79]
	Allylamine	Water, DMF	[80]
	Alkyl amine/amino acid	Dichloromethane (CHCl$_2$), tetrahydrofuran (THF), toluene, dichloromethane (DCM)	[81]
Addition reaction	Polyacetylene	Ortho dichlorobenzene (*O*-DCB)	[82]
	Aryne	DMF, *O*-DCB	[83]
	Poly(oxyalkylenc)amines (POA)	THF	[84]
	Cyclopropanated malonate	Toluene, *O*-DCB, DMF, DCM	[85]

Meanwhile, the non-covalent functionalization method is based on electrostatic interactions such as π–π interactions, van der Waals forces, ionic interactions, and hydrogen bonds [71]; it requires the physical adsorption of suitable molecules on the surface of the graphene. This method does not alter the chemical structure, and it is easy to perform [71]. Non-covalent functionalization is achieved by polymer wrapping, adsorption of surfactants or small aromatic molecules, and interaction with porphyrins or biomolecules.

Non-covalent functionalization is a well-known technique for surface modification of carbon-based nanomaterials. This technique has been previously used to modify the surface of sp^2 networks of CNTs [11, 64]. Current research has shown that the same techniques can be applied with graphene using different types of organic modifiers. Table **3** shows different non-covalent modifications of GO using different modifying agents in various solvents.

Table 3. Non-covalent modification of GO using different modifying agents [64].

Modifying Agent	Dispersing Medium	Dispersibility (mg ml⁻¹)	Electrical Conductivity (S m⁻¹)	Ref.
Amine terminated polymer	1,3-Dimethyl-2-imidazolidinone, γ-butyrolactone, 1-propanol, ethanol, ethylene glycol, DMF	0.4	1500	[86]
Pyrenebutyric acid (PBA)	Water	0.1	200	[87]
poly(sodium 4-styrenesulfonate) (PSS)	Water	1	–	[88]
Coronene derivative	Water	0.15	–	[89]
Sulfonated polyaniline (SPANI)	Water	>1	30	[90]
Porphyrin	Water	0.02	370 Ω cm	[91]
Pyrene-containing hydroxypropyl cellulose (PYR-NHS)	Water	–	–	[92]

The chemical functionalization of graphene and GO offers obvious solutions to the problems associated with graphene. Electron-donating or -withdrawing groups can be bonded to the graphene network by synthetic chemistry methods. These could contribute to the band-gap widening and good dispersibility in common organic solvents. However, while some properties of GO can be improved, the others can decrease using both the covalent and non-covalent methods [71].

Covalent functionalization can be easily performed, and a stronger covalent bond can be formed between GO and the grafted molecules. However, when using this method, the graphene sheets may break down, and their natural conductivity may be compromised. Despite this, the other properties of graphene remain excellent after covalent functionalization [71]. On the contrary, the modification by non-covalent can avoid significant alteration of the structure and therefore not affect important properties of graphene, such as its electrical conductivity and mechanical strength.

CONCLUSION

The extraordinary properties of graphene are becoming very popular. The high numbers of properties of graphene or its derivatives make its spectrum of applications very high, practically limitless, and almost unimaginable to the barriers of human imagination. Various technologies and production methods have been developed for various applications of graphene and its derivatives.

The synthesis routes and derivatives of graphene can be classified into two large categories, which are the bottom-up and top-down strategies. These categories differ according to the synthesized precursor and can encompass all currently known synthesis routes. The synthesis by the bottom-up processes allows greater control, resulting in graphene of higher quality in terms of properties. However, the synthesis of graphene or GO on a large scale is more feasible using top-down processes, such as chemical oxidation-reduction. In this synthetic route, chemical exfoliation by Tour's method illustrates the most suitable method to apply due to its free toxicity and its ability to produce a more organized graphene/GO structure. On the other hand, the thermal reduction method is the most widely used for the reduction of GO to rGO, despite the defects that the structure of graphene can suffer due to high temperatures. Therefore, research efforts are still needed to overcome the hurdles of current synthesis methods, including cost, scalability and environmental concerns, providing new possibilities to achieve improved properties.

Pristine graphene is a hydrophobic material and has no appreciable solubility in most solvents. Graphene processing is primarily concerned with the solubilization of graphene. There are covalent and non-covalent functionalization techniques to improve the solubility of graphene. However, the electrical conductivity of functionalized graphene decreases dramatically compared to unmodified graphene. Furthermore, the surface area of the functionalized graphene prepared by both techniques decreases significantly due to the destructive chemical oxidation of flake graphite followed by sonication, functionalization, and chemical reduction. Several studies have reported the preparation of functionalized graphene from graphite to overcome these drawbacks. In all these cases, the modification of the graphene surface can prevent agglomeration and facilitate the formation of stable dispersions [64].

CONSENT FOR PUBLICATION

Not applicable.

CONFLICT OF INTEREST

The author declares no conflict of interest, financial or otherwise.

ACKNOWLEDGEMENTS

Declared none.

REFERENCES

[1]   Novoselov, K.S.; Jiang, D.; Schedin, F.; Booth, T.J.; Khotkevich, V.V.; Morozov, S.V.; Geim, A.K. Two-dimensional atomic crystals. *Proc. Natl. Acad. Sci. USA,* **2005**, *102*(30), 10451-10453.
[PMID: 16027370]

[2]   Slonczewski, J.C.; Weiss, P.R. Band structure of graphite. *Phys. Rev.,* **1958**, *109*(2), 272-279.

[3]   Sanchez, V.C.; Jachak, A.; Hurt, R.H.; Kane, A.B. Biological interactions of graphene-family nanomaterials: an interdisciplinary review. *Chem. Res. Toxicol.,* **2012**, *25*(1), 15-34.
[PMID: 21954945]

[4]   Ubbelohde, A.R.; Lewis, L.A. *Graphite and its crystal compounds*; Oxford UniversityPress: London, **1960**.

[5]   Boehm, H.P.; Clauss, A.; Fischer, G.; Hofmann, U. Surface properties of extremely thin graphite lamellae. *proceedings of the fifth conference on carbón,* **1962**, pp. 666-669.

[6]   Novoselov, K.S.; Geim, A.K.; Morozov, S.V.; Jiang, D.; Zhang, Y.; Dubonos, S.V.; Grigorieva, I.V.; Firsov, A.A. Electric field effect in atomically thin carbon films. *Science,* **2004**, *306*(5696), 666-669.
[PMID: 15499015]

[7]   Zhang, S.; Yang, K.; Feng, L.; Liu, Z. *In vitro* and *in vivo* behaviors of dextran functionalized graphene. *Carbon,* **2011**, *49*(12), 4040-4049.

[8]   Singh, V.; Joung, D.; Zhai, L.; Das, S.; Khondaker, S.I.; Seal, S. Graphene based materials: Past, present and future. *Prog. Mater. Sci.,* *56*(8), 1178-1271.

[9]   Mohan, V.B.; Lau, K.; Hui, D.; Bhattacharyya, D. Graphene-based materials and their composites: A review on production, applications and product limitations. *Compos. B. Eng.,* **2018**, *142*, 200-220.

[10]   Castro Neto, A.H.; Guinea, F.; Peres, N.M.R.; Novoselov, K.S.; Geim, A.K. The electronic properties of graphene. *Rev. Mod. Phys.,* **2009**, *81*(1), 109-162.

[11]   Kuila, T.; Bose, S.; Hong, C.E.; Uddin, M.E.; Khanra, P.; Kim, N.H. Preparation of functionalized graphene/linear low density polyethylene composites by a solution mixing method. *Carbon,* **2011**, *49*(3), 1033-1037.

[12]   Fan, H.; Wang, L.; Zhao, K.; Li, N.; Shi, Z.; Ge, Z.; Jin, Z. Fabrication, mechanical properties, and biocompatibility of graphene-reinforced chitosan composites. *Biomacromolecules,* **2010**, *11*(9), 2345-2351.
[PMID: 20687549]

[13]   Ansari, S.; Giannelis, E.P. Functionalized graphene sheet-poly (vinylidene fluoride) conductive nanocomposites. *J. Polym. Sci., B, Polym. Phys.,* **2009**, *47*(9), 888-897.

[14]   Park, H.; Rowehl, J.A.; Kim, K.K.; Bulovic, V.; Kong, J. Doped graphene electrodes for organic solar cells. *Nanotechnology,* **2010**, *21*(50), 505204.
[PMID: 21098945]

[15]   Wu, J.; Becerril, H.A.; Bao, Z.; Liu, Z.; Chen, Y.; Peumans, P. Organic solar cells with solution-processed graphene transparent electrodes. *Appl. Phys. Lett.,* **2008**, *92*(26), 263302.

[16]   Prasai, D.; Tuberquia, J.C.; Harl, R.R.; Jennings, G.K.; Rogers, B.R.; Bolotin, K.I. Graphene: corrosion-inhibiting coating. *ACS Nano,* **2012**, *6*(2), 1102-1108.
[PMID: 22299572]

[17]   Lai, W.F.; Wong, W.T. Design of polymeric gene carriers for effective intracellular delivery. *Trends Biotechnol.,* **2018**, *36*(7), 713-728.
[PMID: 29525137]

[18]   Lai, W.F.; Green, D.W.; Jung, H.S. Linear poly(ethylenimine) cross-linked by methyl-β-cyclodextrin for gene delivery. *Curr. Gene Ther.,* **2014**, *14*(4), 258-268.
[PMID: 25039611]

[19] Shi, X.; Zheng, S.; Wu, Z.S.; Bao, X. Recent advances of graphene-based materials for high-performance and new-concept supercapacitors. *J. Energy Chem.,* **2018**, *27*(1), 25-42.

[20] Solís-Fernández, P.; Bissett, M.; Ago, H. Synthesis, structure and applications of graphene-based 2D heterostructures. *Chem. Soc. Rev.,* **2017**, *46*(15), 4572-4613.
[PMID: 28691726]

[21] Hernaez, M.; Zamarreño, C.R.; Melendi-Espina, S.; Bird, L.R.; Mayes, A.G.; Arregui, F.J. Optical Fibre Sensors Using Graphene-Based Materials: A Review. *Sensors (Basel),* **2017**, *17*(1), 155.
[PMID: 28098825]

[22] Yang, K.; Wang, J.; Chen, X.; Zhao, Q.; Ghaffar, A.; Chen, B. Application of graphene-based materials in water purification: from the nanoscale to specific devices. *Environ. Sci. Nano,* **2018**, *5*(6), 1264-1297.

[23] Perreault, F.; Fonseca de Faria, A.; Elimelech, M. Environmental applications of graphene-based nanomaterials. *Chem. Soc. Rev.,* **2015**, *44*(16), 5861-5896.
[PMID: 25812036]

[24] Zurutuza, A.; Marinelli, C. Challenges and opportunities in graphene commercialization. *Nat. Nanotechnol.,* **2014**, *9*(10), 730-734.
[PMID: 25286257]

[25] Ahmed, F.; Rodrigues, D.F. Investigation of acute effects of graphene oxide on wastewater microbial community: a case study. *J. Hazard. Mater.,* **2013**, *256-257*, 33-39.
[PMID: 23669788]

[26] Zhen, Z.; Zhu, H. Structure and Properties of Graphene. *Graphene.,* **2018**, 1-12.

[27] Yang, G.; Li, L.; Lee, W.B.; Ng, M.C. Structure of graphene and its disorders: a review. *Sci. Technol. Adv. Mater.,* **2018**, *19*(1), 613-648.
[PMID: 30181789]

[28] Guo, H.; Liu, R.; Cheng, Z.X.; Wu, X. *Graphene-Based Architecture and Assemblies*; Graphene Chemistry, **2013**, pp. 153-182.

[29] Katsnelson, M.I. *Graphene: carbon in two dimensions*; Cambridge University Press: Cambridge, **2012**.

[30] Nemes-Incze, P.; Osváth, Z.; Kamarás, K.; Biró, L.P. Anomalies in thickness measurements of graphene and few layer graphite crystals by tapping mode atomic force microscopy. *Carbon,* **2008**, *46*(11), 1435-1442.

[31] Choi, W.; Lee, J. Graphene: Synthesis and Applications, CRC Press, Taylor and Francis Group; Boca Raton London New York, Chapter –2 , 27-57.

[32] Gupta, A.; Chen, G.; Joshi, P.; Tadigadapa, S.; Eklund, P.C. Raman scattering from high-frequency phonons in supported n-graphene layer films. *Nano Lett.,* **2006**, *6*(12), 2667-2673.
[PMID: 17163685]

[33] Kauling, A.P.; Seefeldt, A.T.; Pisoni, D.P.; Pradeep, R.C.; Bentini, R.; Oliveira, R.V.B.; Novoselov, K.S.; Castro Neto, A.H. The Worldwide Graphene Flake Production. *Adv. Mater.,* **2018**, *30*(44), e1803784.
[PMID: 30209839]

[34] Bianco, A.; Cheng, H.M.; Enoki, T.; Gogotsi, Y.; Hurt, R.H.; Koratkar, N.; Zhang, J. All in the graphene family – A recommended nomenclature for two-dimensional carbon materials. *Carbon,* **2013**, *65*, 1-6.

[35] Kumar, N.; Salehiyan, R.; Chauke, V.; Botlhoko, O.J.; Setshedi, K.; Scriba, M.; Ray, S.S. Top-down synthesis of graphene: A comprehensive review. Flatchem, 100224 [36] Neto, A. C.; Guinea, F.; Peres, N.M. Drawing conclusions from graphene. *Phys. World,* **2006**, *19*(11), 33-37.

[36]   Cheng, C.; Li, S.; Thomas, A.; Kotov, N.A.; Haag, R. Functional Graphene Nanomaterials Based Architectures: Biointeractions, Fabrications, and Emerging Biological Applications. *Chem. Rev.,* **2017,** *117*(3), 1826-1914.
[PMID: 28075573]

[37]   Park, S.; Ruoff, R.S. Chemical methods for the production of graphenes. *Nat. Nanotechnol.,* **2009,** *4*(4), 217-224.
[PMID: 19350030]

[38]   Lim, J.Y.; Mubarak, N.M.; Abdullah, E.C.; Nizamuddin, S.; Khalid, M. Recent trends in the synthesis of graphene and graphene oxide based nanomaterials for removal of heavy metals–A review. *Ind. Eng. Chem. Res.,* **2018,** *66*, 29-44.

[39]   Muñoz, R.; Gómez-Aleixandre, C. Review of CVD synthesis of graphene. *Chem. Vapor Depos,* **2013,** *19*(10-11-12), 297-322.

[40]   Tetlow, H.; De Boer, J.P.; Ford, I.J.; Vedensky, D.D.; Coraux, J.; Kantorovich, L. Growth of epitaxial graphene: Theory and experiment. *Phys. Rep.,* **2014,** *542*(3), 195-295.

[41]   Li, X.; Cai, W.; An, J.; Kim, S.; Nah, J.; Yang, D.; Piner, R.; Velamakanni, A.; Jung, I.; Tutuc, E.; Banerjee, S.K.; Colombo, L.; Ruoff, R.S. Large-area synthesis of high-quality and uniform graphene films on copper foils. *Science,* **2009,** *324*(5932), 1312-1314.
[PMID: 19423775]

[42]   Rümmeli, M.H.; Bachmatiuk, A.; Scott, A.; Börrnert, F.; Warner, J.H.; Hoffman, V.; Lin, J.H.; Cuniberti, G.; Büchner, B. Direct low-temperature nanographene CVD synthesis over a dielectric insulator. *ACS Nano,* **2010,** *4*(7), 4206-4210.
[PMID: 20586480]

[43]   Chaste, J.; Saadani, A.; Jaffre, A.; Madouri, A.; Alvarez, J.; Pierucci, D.; Ouerghi, A. Nanostructures in suspended mono-and bilayer epitaxial graphene. *Carbon,* **2017,** *125*, 162-167.

[44]   Yu, X.; Liu, J.; Xiao, C. Numerical computation of flow and hydrodynamic force of ellipsoid based on DES hybrid methods. *J Naval Uni Eng,* **2011,** *23*(1), 100-103.

[45]   Somani, P.R.; Somani, S.P.; Umeno, M. Planer nano-graphenes from camphor by CVD. *Chem. Phys. Lett.,* **2006,** *430*(1-3), 56-59.

[46]   Bhuyan, M.S.A.; Uddin, M.N.; Islam, M.M.; Bipasha, F.A.; Hossain, S.S. Synthesis of graphene. *Int. Nano Lett.,* **2016,** *6*(2), 65-83.

[47]   Xiong, W.; Zhou, Y.S.; Hou, W.J.; Jiang, L.J.; Gao, Y.; Fan, L.S.; Jiang, L.; Silvain, J.F.; Lu, Y.F. Direct writing of graphene patterns on insulating substrates under ambient conditions. *Sci. Rep.,* **2014,** *4*(1), 4892.
[PMID: 24809639]

[48]   Ye, R.; James, D.K.; Tour, J.M. Laser-Induced Graphene. *Acc. Chem. Res.,* **2018,** *51*(7), 1609-1620.
[PMID: 29924584]

[49]   Subrahmanyam, K.S.; Panchakarla, L.S.; Govindaraj, A.; Rao, C.N.R. Simple method of preparing graphene flakes by an arc-discharge method. *J. Phys. Chem. C,* **2009,** *113*(11), 4257-4259.

[50]   Song, Z.; Mu, X.; Luo, T.; Xu, Z. Unzipping of carbon nanotubes is geometry-dependent. *Nanotechnology,* **2016,** *27*(1), 015601.
[PMID: 26597779]

[51]   Amiri, A.; Naraghi, M.; Ahmadi, G.; Soleymaniha, M.; Shanbedi, M. A review on liquid-phase exfoliation for scalable production of pure graphene, wrinkled, crumpled and functionalized graphene and challenges. *FlatChem,* **2018,** *8*, 40-71.

[52]   Hernandez, Y.; Nicolosi, V.; Lotya, M.; Blighe, F.M.; Sun, Z.; De, S.; McGovern, I.T.; Holland, B.; Byrne, M.; Gun'Ko, Y.K.; Boland, J.J.; Niraj, P.; Duesberg, G.; Krishnamurthy, S.; Goodhue, R.; Hutchison, J.; Scardaci, V.; Ferrari, A.C.; Coleman, J.N. High-yield production of graphene by liquid-

phase exfoliation of graphite. *Nat. Nanotechnol.,* **2008**, *3*(9), 563-568.
[PMID: 18772919]

[53]   Zhang, Y.B.; Small, J.P.; Pontius, W.V.; Kim, P. Fabrication and electric-field dependent transport measurements of mesoscopic graphite devices. *Appl. Phys. Lett.,* **2005**, *86*, 073104.

[54]   Yu, P.; Lowe, S.E.; Simon, G.P.; Zhong, Y.L. Electrochemical exfoliation of graphite and production of functional graphene. *Curr. Opin. Colloid Interface Sci.,* **2015**, *20*(5-6), 329-338.

[55]   Saba, N.; Jawaid, M.; Fouad, H.; Alothman, O.Y. Nanocarbon: Preparation, properties, and applications. Nanocarbon and its. *Composites,* **2019**, 327-354.

[56]   Morozov, S.V.; Novoselov, K.S.; Katsnelson, M.I.; Schedin, F.; Elias, D.C.; Jaszczak, J.A.; Geim, A.K. Giant intrinsic carrier mobilities in graphene and its bilayer. *Phys. Rev. Lett.,* **2008**, *100*(1), 016602.
[PMID: 18232798]

[57]   Huang, X.; Qi, X.; Boey, F.; Zhang, H. Graphene-based composites. *Chem. Soc. Rev.,* **2012**, *41*(2), 666-686.
[PMID: 21796314]

[58]   Zhou, L.; Wang, Y.; Cao, G. Estimating the elastic properties of few-layer graphene from the free-standing indentation response. *J. Phys. Condens. Matter,* **2013**, *25*(47), 475301.
[PMID: 24166876]

[59]   Kudin, K.N.; Scuseria, G.E.; Yakobson, B.I.Y. C2F, BN, and C nanoshell elasticity fromab initiocomputations. *Phys. Rev. B Condens. Matter Mater. Phys.,* **2001**, *64*(23)

[60]   Lee, C.; Wei, X.; Kysar, J.W.; Hone, J. Measurement of the elastic properties and intrinsic strength of monolayer graphene. *Science,* **2008**, *321*(5887), 385-388.
[PMID: 18635798]

[61]   Faugeras, C.; Faugeras, B.; Orlita, M.; Potemski, M.; Nair, R.R.; Geim, A.K. Thermal conductivity of graphene in corbino membrane geometry. *ACS Nano,* **2010**, *4*(4), 1889-1892.
[PMID: 20218666]

[62]   Gadipelli, S.; Guo, Z.X. Graphene-based materials: Synthesis and gas sorption, storage and separation. Progress in J. *Mater. Sci.,* **2015**, *69*, 1-60.

[63]   Balandin, A.A.; Ghosh, S.; Bao, W.; Calizo, I.; Teweldebrhan, D.; Miao, F.; Lau, C.N. Superior thermal conductivity of single-layer graphene. *Nano Lett.,* **2008**, *8*(3), 902-907.
[PMID: 18284217]

[64]   Kuila, T.; Bose, S.; Mishra, A.K.; Khanra, P.; Kim, N.H.; Lee, J.H. Chemical functionalization of graphene and its applications. *Prog. Mater. Sci.,* **2012**, *57*(7), 1061-1105.

[65]   Sun, Z.; Yan, Z.; Yao, J.; Beitler, E.; Zhu, Y.; Tour, J.M. Growth of graphene from solid carbon sources. *Nature,* **2010**, *468*(7323), 549-552.
[PMID: 21068724]

[66]   Zhang, P.; Ma, L.; Fan, F.; Zeng, Z.; Peng, C.; Loya, P.E.; Liu, Z.; Gong, Y.; Zhang, J.; Zhang, X.; Ajayan, P.M.; Zhu, T.; Lou, J. Fracture toughness of graphene. *Nat. Commun.,* **2014**, *5*(1), 3782.
[PMID: 24777167]

[67]   Foo, M.E.; Gopinath, S.C.B. Feasibility of graphene in biomedical applications. *Biomed. Pharmacother.,* **2017**, *94*, 354-361.
[PMID: 28772213]

[68]   Loh, K.P.; Bao, Q.; Ang, P.K.; Yang, J. The chemistry of graphene. *J. Mater. Chem.,* **2010**, *20*(12), 2277.

[69]   Wu, Y.; Lin, Y.M.; Bol, A.A.; Jenkins, K.A.; Xia, F.; Farmer, D.B.; Zhu, Y.; Avouris, P. High-frequency, scaled graphene transistors on diamond-like carbon. *Nature,* **2011**, *472*(7341), 74-78.
[PMID: 21475197]

[70] Yang, G.; Bao, D.; Liu, H.; Zhang, D.; Wang, N.; Li, H. Functionalization of Graphene and Applications of the Derivatives. *J. Inorg. Organomet. Polym. Mater.,* **2017**, *27*(5), 1129-1141.

[71] Ismail, A.N.; Zulkifli, N.W.M.; Chowdhury, Z.Z.; Johan, M.F. Functionalization of graphene-based materials: Effective approach for enhancement of tribological performance as lubricant additives. *Diamond Related Materials,* **2021**, *115*, 08357.

[72] Salavagione, H.J.; Gómez, M.A.; Martínez, G. Polymeric Modification of Graphene through Esterification of Graphite Oxide and Poly (vinyl alcohol). *Macromolecules,* **2009**, *42*(17), 6331-6334.

[73] Shen, J.; Yan, B.; Shi, M.; Ma, H.; Li, N.; Ye, M. Synthesis of graphene oxide-based biocomposites through diimide-activated amidation. *J. Colloid Interface Sci.,* **2011**, *356*(2), 543-549.
      [PMID: 21329939]

[74] Worsley, K.A.; Ramesh, P.; Mandal, S.K.; Niyogi, S.; Itkis, M.E.; Haddon, R.C. Soluble graphene derived from graphite fluoride. *Chem. Phys. Lett.,* **2007**, *445*(1-3), 51-56.

[75] Niyogi, S.; Bekyarova, E.; Itkis, M.E.; McWilliams, J.L.; Hamon, M.A.; Haddon, R.C. Solution properties of graphite and graphene. *J. Am. Chem. Soc.,* **2006**, *128*(24), 7720-7721.
      [PMID: 16771469]

[76] Stankovich, S.; Piner, R.D.; Nguyen, S.T.; Ruoff, R.S. Synthesis and exfoliation of isocyanate-treated graphene oxide nanoplatelets. *Carbon,* **2006**, *44*(15), 3342-3347.

[77] Sun, Z.; Kohama, S.; Zhang, Z.; Lomeda, J.R.; Tour, J.M. Soluble graphene through edge-selective functionalization. *Nano Res.,* **2010**, *3*(2), 117-125.

[78] Pham, T.A.; Kumar, N.A.; Jeong, Y.T. Covalent functionalization of graphene oxide with polyglycerol and their use as templates for anchoring magnetic nanoparticles. *Synth. Met.,* **2010**, *160*(17-18), 2028-2036.

[79] Kuila, T.; Khanra, P.; Bose, S.; Kim, N.H.; Ku, B.C.; Moon, B.; Lee, J.H. Preparation of water-dispersible graphene by facile surface modification of graphite oxide. *Nanotechnology,* **2011**, *22*(30), 305710.
      [PMID: 21730750]

[80] Wang, G.; Wang, B.; Park, J.; Yang, J.; Shen, X.; Yao, J. Synthesis of enhanced hydrophilic and hydrophobic graphene oxide nanosheets by a solvothermal method. *Carbon,* **2009**, *47*(1), 68-72.

[81] Bourlinos, A.B.; Gournis, D.; Petridis, D.; Szabó, T.; Szeri, A.; Dékány, I. Graphite Oxide: Chemical Reduction to Graphite and Surface Modification with Primary Aliphatic Amines and Amino Acids. *Langmuir,* **2003**, *19*(15), 6050-6055.

[82] Xu, X.; Luo, Q.; Lv, W.; Dong, Y.; Lin, Y.; Yang, Q.; Li, Z. Functionalization of Graphene Sheets by Polyacetylene: Convenient Synthesis and Enhanced Emission. *Macromol. Chem. Phys.,* **2011**, *212*(8), 768-773.

[83] Zhong, X.; Jin, J.; Li, S.; Niu, Z.; Hu, W.; Li, R.; Ma, J. Aryne cycloaddition: highly efficient chemical modification of graphene. *Chem. Commun. (Camb.),* **2010**, *46*(39), 7340-7342.
      [PMID: 20820532]

[84] Hsiao, M.C.; Liao, S.H.; Yen, M.Y.; Liu, P.I.; Pu, N.W.; Wang, C.A.; Ma, C.C.M. Preparation of covalently functionalized graphene using residual oxygen-containing functional groups. *ACS Appl. Mater. Interfaces,* **2010**, *2*(11), 3092-3099.
      [PMID: 20949901]

[85] Economopoulos, S.P.; Rotas, G.; Miyata, Y.; Shinohara, H.; Tagmatarchis, N. Exfoliation and chemical modification using microwave irradiation affording highly functionalized graphene. *ACS Nano,* **2010**, *4*(12), 7499-7507.
      [PMID: 21080708]

[86] Choi, E.Y.; Han, T.H.; Hong, J.; Kim, J.E.; Lee, S.H.; Kim, H.W.; Kim, S.O. Noncovalent functionalization of graphene with end-functional polymers. *J. Mater. Chem.,* **2010**, *20*(10), 1907.

[87]   Xu, Y.; Bai, H.; Lu, G.; Li, C.; Shi, G. Flexible graphene films *via* the filtration of water-soluble noncovalent functionalized graphene sheets. *J. Am. Chem. Soc.,* **2008**, *130*(18), 5856-5857. [PMID: 18399634]

[88]   Stankovich, S.; Piner, R.D.; Chen, X.; Wu, N.; Nguyen, S.T.; Ruoff, R.S. Stable aqueous dispersions of graphitic nanoplatelets *via* the reduction of exfoliated graphite oxide in the presence of poly(sodium 4-styrenesulfonate). *J. Mater. Chem.,* **2006**, *16*(2), 155-158.

[89]   Ghosh, A.; Rao, K.V.; George, S.J.; Rao, C.N.R. Noncovalent functionalization, exfoliation, and solubilization of graphene in water by employing a fluorescent coronene carboxylate. *Chemistry,* **2010**, *16*(9), 2700-2704. [PMID: 20108284]

[90]   Bai, H.; Xu, Y.; Zhao, L.; Li, C.; Shi, G. Non-covalent functionalization of graphene sheets by sulfonated polyaniline. *Chem. Commun. (Camb.),* **2009**, (13), 1667-1669. [PMID: 19294256]

[91]   Geng, J.; Jung, H.T. Porphyrin Functionalized Graphene Sheets in Aqueous Suspensions: From the Preparation of Graphene Sheets to Highly Conductive Graphene Films. *J. Phys. Chem.,* **2010**, *114*(18), 8227-8234.

[92]   Kodali, V.K.; Scrimgeour, J.; Kim, S.; Hankinson, J.H.; Carroll, K.M.; de Heer, W.A.; Berger, C.; Curtis, J.E. Nonperturbative chemical modification of graphene for protein micropatterning. *Langmuir,* **2011**, *27*(3), 863-865. [PMID: 21182241]

[93]   Chang, H.; Wang, G.; Yang, A.; Tao, X.; Liu, X.; Shen, Y.; Zheng, Z.A. Transparent, Flexible, Low-Temperature, and Solution-Processible Graphene Composite Electrode. *Adv. Funct. Mater.,* **2010**, *20*(17), 2893-2902.

Graphene-based Nanomaterials as Organocatalyst

Angeliki Brouzgou[1,*]

[1] *Department of Energy Systems, School of Technology, University of Thessaly, Geapolis, Regional Road Trikala-Larisa, 41500 Larisa, Greece*

Abstract: Graphene is a π-electron rich material capable of interacting with organic molecules. Its derivative graphene oxide (or reduced graphene oxide) offers a large surface area, many hydroxyl, carboxyl and epoxy groups which can accommodate organic molecules that act as active catalytic sites. The graphene oxide supported organocatalysts have boosted significantly the organocatalytic reactions increasing more than ninety percent the catalytic activity and selectivity of specific reactions to important industrial products. The graphene-based heterogeneous organocatalysts are stable, recyclable and of lower cost. The functionalization of graphene oxide with acids, the synthesis of bifunctional acid-base graphene-based catalysts and the doping of graphene with non-metal elements, are the main strategies that are followed for the synthesis of GO-based heterogeneous organocatalysts.

Keywords: Acid-base graphene oxide, Amines functionalization, Asymmetric catalysis, Bifunctional catalyst, Chiral catalysts, Co-doping graphene, Electro-organic synthesis, Electro-organic transformation, Functionalization, Grafting graphene-based materials, Graphene, Graphene oxide structure, Graphene oxide, Heterogeneous catalysis, Immobilization, Magnetic nanoparticles, Organic-catalysis, Organic molecule, Recyclable graphene catalyst, Reduced graphene oxide.

INTRODUCTION

Heterogeneous electro-organic catalysts developed in the last few decades paved the way for organocatalysis to pass over from homogeneous to heterogeneous phase, boosting many industrial processes, especially in the pharmaceutical and food industry. The scientists, after many years of research, managed the replacement of the metal with an organic substrate, leading to many imperative reactions that are included in the manufacturing of the daily product, from enviro-

* **Corresponding author Angeliki Brouzgou:** Department of Energy Systems, School of Technology, University of Thessaly, Geapolis, Regional Road Trikala-Larisa, 41500 Larisa, Greece; E-mail: brouzgou@gmail.com

Manorama Singh, Vijai K. Rai and Ankita Rai (Eds)

nmental unfriendly, time consuming, of high cost as well as of low yield, to the opposite direction [1].

Explicitly, the integration of electrochemical oxidation into chiral amine catalysis, with the chiral primary amine to the role of catalyst, presented excellent enantioselectivity as well as high yields [2]. The economic impact of such performances is huge and so there is a huge demand for further progress of such reactions, giving 100% yield, using recyclable catalysts, fulfilling all the desired characteristics, as well [3].

Graphene-based organocatalysts have caused great anticipation as graphene 2D structure and its outstanding mechanical and electric characteristics in tandem with the intrinsic functional moieties and those that can accommodate onto its surface, show impressive catalytic properties for many organic synthesis and formation reactions and high recyclability, as well.

As the literature works state, the hybrid graphene-based supported heterogeneous organocatalysts are the key concept for achieving the desired characteristics of an electroorganic catalyst, and in a sustainable way as well. This might seem simple, however, the scientific community should further recognize the interactions between the support and the organic molecule, the covalent and non-covalent structure reactivity, also identifying many other influential factors [3].

ACID FUNCTIONALIZED GRAPHENE OXIDE AND REDUCED GRAPHENE OXIDE NANOCOMPOSITES

Chiral catalysts that are natural nitrogen sources, such as alkaloids, have been explored extensively into organocatalytic reactions [4]. A chiral substance usually acts as Lewis base [5] that enhances the enantioselectivity, however under electrochemical oxidation processes is oxidized itself [6]. This issue motivated many research groups to place organocatalysts onto solid supports, achieving higher yields, stability, recyclability and reusability of them.

The covalent or non-covalent bonding between the organocatalysts and the support proffered recoverable materials with multifunctional surface [7]. The special structure of graphene material and its derivatives due to its defects and functionalities have been applied successfully for many reactions of great interest [8 - 10] in various fields. However, the use of graphene material (and its derivatives) as support for the organocatalysts immobilization is still an immature field that leaves room for more future investigation.

The functionalization of graphene oxide or reduced graphene oxide with amino acids is a nexus creating covalent bonds between them and so delivering new functional groups and carboxylic acid groups that are advantageous for the progress of the reaction of interest. The covalent bonding between the organic molecule can take place through hydrogen bond interactions and the graphene material provides a more stable catalyst as well as an easily recyclable one [11].

L-proline covalently anchored onto graphene oxide skeleton signifies an important strategy in organocatalysis field for creating functional groups. From the first years of electroorganic catalysis, L-proline has been extensively investigated and applied in many organocatalytic reactions, with the direct asymmetric aldol one to be the most sought after, especially for the pharmaceutical industry [11, 12].

Graphene sheets, even after the functionalization process, must retain their initial homogeneous morphology, and so the process of functionalization can affect the heterogeneous catalyst efficiency. Interestingly, also the type of solvent (polar or non-polar) affects the graphene-based catalyst maximum yield as well as stereoselectivity.

Generally, the functionalization of GO and rGO can alert a hydrophobic nature of them and so the most effective performance can be obtained into nonpolar solvents [12]. Additionally, the electronic nature of the aromatic aldehydes can be responsible for the maximum obtained yield of organocatalysts; while the poor electron ones (nitrobenzaldehydes, cyanobenzaldehyde and (trifluoromethyl)benzaldehyde) can give better yields.

Furthermore, trans-4-hydroxy-l-proline supported onto graphene oxide [13] was achieved with almost 40% immobilization efficiency. The covalent bonding is accomplished between the hydroxyl groups of the organic molecules and the carboxylic acid (-COOH) of GO. During the functionalization procedure, initially, the GO carbonylation (due to GO reduction) takes place and then the L-proline molecules are embedded into GO structure, increasing its mechanical strain, while the chemical modification follows. The homogeneous distribution of the basic elements, C, O and N, onto the GO surface indicates the uniform attachment of the functional material onto the support [13]. This is a crucial factor for attaining a high organo-catalytic reactions yield, as both sides of the 2D organic molecules encounter the reactants and products, reducing the mass transfer resistance.

Except for L-proline, many other linear amino acids have been examined for graphene sheets covalent functionalization. The research group of Sadiq *et al.* [14] suggested a three-step functionalization strategy, according to which the covalent bond is created between the amino acids -COOH groups, not the -NH$_2$ groups as commonly followed, and the graphene sheets (Fig. **1**). The -COOH groups are formed during the graphene oxidation, and then they are converted into -COCl groups, while the treatment with BOC-amino acid assures that the -COOH and not the -NH$_2$ groups will create the covalent bond. The resulted hybrid heterocatalyst presented a high yield, almost 85%, for the asymmetric aldol reactions of aldehydes and ketones through emine route with the water to be the only solvent [14].

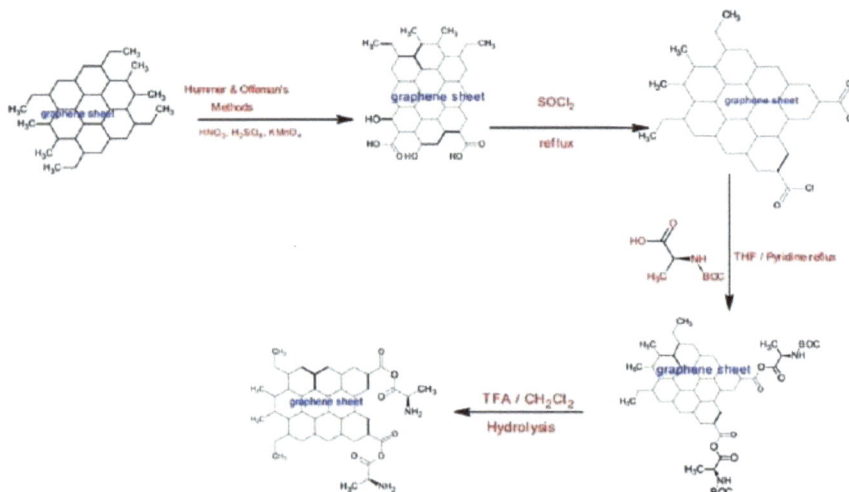

Fig. (1). A suggested synthesis for covalently immobilized amines onto GO sheets [14].

Therefore, the proline, its derivatives and other amino acids, due to their facile use for functionalization, paved the way for the synthesis of supported heterogenous organocatalysts. Even though the as-synthesized novel materials increased the yield of the reactions providing also the possibility of recyclability, some catalytic structures demonstrated mass loss, and were poisoned by organic impurities [12]. While in some other cases, a small percentage of L-proline transmuted to other substances was observed [13]. As Zhou *et al.* [15] stated, also their limited BET surface and the weak interaction of the support and the organic molecule can result in undesired leaching. In order to overcome those arisen challenges, the covalent organic frameworks (COFs) come to the fore as more promising materials for heterogeneous organocatalysts [16 - 20]. The novel COFs materials were fabricated in 2005 from a group of researchers [21] and since then they have been mainly investigated into energy storage systems and for Michael addition reactions [22, 23]. Their organic framework is reported to be like graphene, while their porous crystalline structure can act as an organic platform for facilitating reactions or as a catalyst itself [24, 25]. Moreover, when they are altered with chiral catalytic moieties, enantioselective products can be formed [25]. As also being notified, the optimum crystallinity, stability and porosity of COFs can be achieved in one covalent framework only if the interlayer interactions are the appropriate ones. Therefore, the development of 2D or 3D COFs on thin film substrates, such as graphene, is one simple promising method for improving COFs' performance. A COF/single layer graphene can be synthesized by a simple solvothermal method [26] which is the most common synthesis strategy. However, the substrate adoption creates a new challenge; the difficulty of the control of the thickness layer [27, 28].

Graphene as a complex material with various edge states and defects may cause different interactions facilitating an on-site condensation reaction enabling the synthesis of COF/graphene material. The synthesis of 2D-COF/single layer graphene nanohybrid catalysts *via* alternative and greener processes is a future promise from the research community [29 - 31].

A few works are found in literature that have examined GO functionalization with acids, such as sulfanilic acid and citric acid. The sulfanilic acids functionalized graphene oxide or reduced graphene oxide are considered good electroorganic catalysts for acid-based organic transformations but still have not been explored in depth by the research community. The covalent bonding between sulfanilic acids and the graphene sheet is succeeded *via* the initial production of diazonium salts and the immediate transfer of graphene's electron to the unstable aryl radical that is formed from the unstable sulfanilic acids that lost their nitrogen [32].

The GO functionalization with citric acid is recently reported for the first time by Maleki *et al.* [33]. Citric acid is an eco-friendly organic substance that has plenty of carboxylic groups which when immobilized onto the GO surface, provide a heterogeneous catalyst which can successfully catalyze the synthesis of imidazole derivatives 3a-n *via* the reaction of benzaldehyde 1 and o-phenylenediamine 2.

The covalent functionalization of the GO material structure with amines and hydroxyl produces stable heterogenous GO-based organocatalysts in terms of catalytic activity, recyclability, and structure when they are suspended into a solvent.

ACID-BASE BIFUNCTIONAL GRAPHENE OXIDE

There are many electroorganic reactions that demand the presence of acidic and basic catalysts, meaning the activation of electrophiles and nucleophiles, in order to process [34]. However, the presence of both acidic and basic catalysts into a reactor is problematic and so the research community suggests the concept of 'site isolation', mimicking the enzymatic catalysis; according to which

an enzyme does not catalyze undesired interactions with the aid of incompatible sites that are randomly distributed. Graphene oxide as a material contains carboxylic acid groups due to which owns acidic characteristics. The research groups exploiting this GO's acidic nature managed to synthesize acid-base bifunctional heterogeneous organocatalysts (Fig. **2A**) [35].

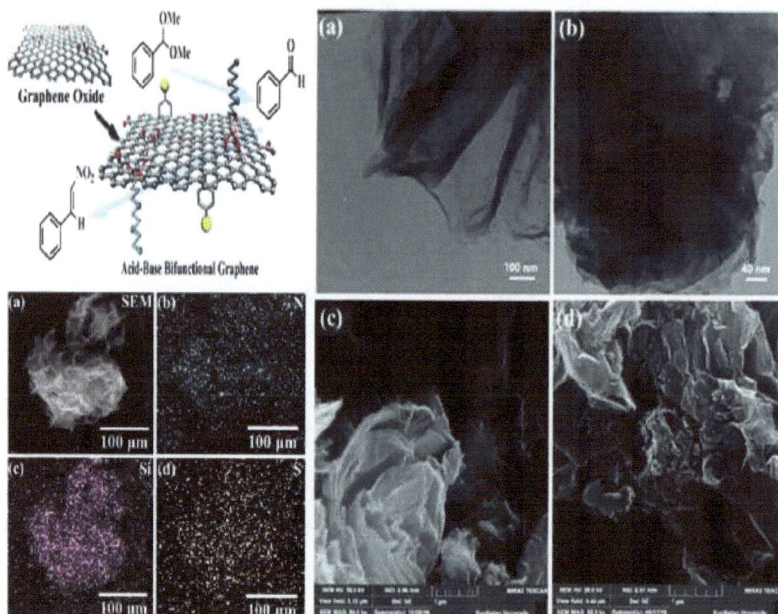

Fig. (2). Acid-base bifunctional graphene (top) and the respective SEM images (**a**, bottom) (bottom) and EDS element analysis (**b**, **c** and **d**, bottom)(A) [35]; TEM (**a**, **b**) and SEM (**c,d**) images of piperazine-GO hybrid catalyst [36].

The grafted basic groups onto the GO plane and the carboxylic acid groups act synergistically, providing excellent catalytic activity and recyclability as well as reusability. Furthermore, the use of natural materials such as piperazine for acid-base bifunctional graphene organocatalysts make them even more attractive, greener and sustainable materials. Fig. (**2B**) shows the morphological features of piperazine-GO heterogeneous catalyst [37].

Enriching the GO (or rGO) surface with electrons, acquires the ability to stabilize the cationic intermediates that are produced during an electroorganic reaction [38, 39]. According to the anchoring process that has been followed by many research groups [38, 39], after the GO or rGO production with the known modified Hummer's method, the piperazine is added [36, 40], in order the ring epoxides to open *via* the nucleophilic substitution process; and then follows a reduction process with hydrazine. Fig. (**3**) compares the iminium activation and basic activation for the Knoevenagel and Michael reactions. The electron rich GO surface, which acts as a Lewis base, can immobilize the iminium intermediate and to promote the catalytic reaction, indications its (acidic-basic) bifunctional role.

Fig. (3). Forms of catalytic activation (top) and high nitrogen content increase (bottom) [39].

Ramírez-Jiménez *et al.* [39] also investigated the effect of various functionalized alcohols on the opening process of epoxides. They found out that the 1,8-diazabicycloundec-7-ene (DBU) and N,N-dimethylformamide (DMF) except for the S -inclusion into the GO matrix, can significantly increase the nitrogen content (Fig. 3). Furthermore, according to their observations, when the DBU and DMF react with each other amine 1 and $HNMe_2$ are produced that are covalently bonding onto the GO surface. Additionally, the use of DMF as the mean for the functionalized GO reduction process could lead to the development of a greener synthesis method, excluding hydrazine from it.

Pyridine grafted GO heterogeneous catalysts are also reported in the literature [41] as acid-base bifunctional catalysts for one-pot synthesis of b-phosphonomalonates *via* cascade Knoevenagel–phospha Michael addition reaction in water. The homogeneously distributed active sites and the folded structure reduce the mass transfer resistance and so the catalytic performance increases (Fig. 4).

It is remarked that the rGO, in contrast to the GO, when is subjected to reduction process (thermal or chemical), possesses hydrophobic behavior [42], making its dispersion into a polymer matrix a challenging procedure. Thus, the rGO surface modification with, for instance pyridine groups, also amends dispersion of rGO into polar solvents [43].

Recently, Choudhury *et al.* [44] achieved a high yield selective formation n of 1,4-dihy functionalized 1,4-dihydropyridines (DHP), acridinediones and polyhydroquinolines with the aid of amine-functionalized GO nanosheets, under mild conditions. According to the authors, the high and selective catalytic performance was attributed to an observed association effect between the GO carboxylic groups and the anchored onto its surface amines. This effect made the as-synthesized catalyst, recyclable without presenting any mass loss even after five consecutive measurements.

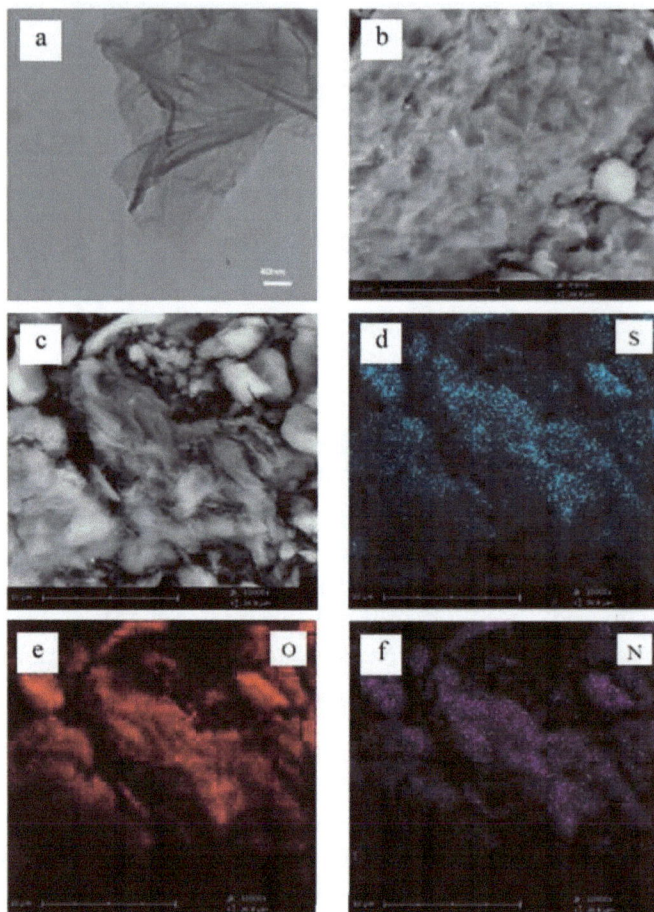

Fig. (4). TEM (**a**), SEM (**b, c e**) and EDS element analysis of pyridine grafted GO sheets [41].

METAL-FREE DOPED GRAPHENE

Doping a carbon source material with an N,P or another non-metal element in order to improve its catalytic properties has been proved to be a beneficial method for many electrocatalytic reactions [45, 46]. The introduction of a non-metal element into graphene lattice [37, 47 - 49] alters its both electronic and chemical features, and so making graphene even more suitable for organocatalytic reactions [50, 51].

Graphene (especially pristine graphene) which is unstable and quite passive as organocatalyst, when is doped with an element with different electronegativities, such as N, B, S, P, *etc.*, acquires charged sites that favor the adsorption of more reactants as well as the catalytic performance. For instance, the introduction of N atoms into graphene sheets urges to the carbon atoms that they adapt to, local positive charge and spin density, that is similar to metallic features, and so they have been investigated for various metal-free organic transformations [52].

The N-doped graphene materials are also being reported to own higher catalytic surface area and altered energy band gap that are two desired characteristics for promoting redox processes. In literature, most of the research groups have investigated N-doped graphene organocatalysts for aromatic compounds reactions, such as nitrophenol reduction [53], ethylbenzene selective oxidation [54] and others [55]. Recently, it is proved that N-doped graphene materials are very effective for N-formylation of amines with CO_2 for formamide production [56], exhibiting yields higher than 99%, presenting excellent recyclability, as well.

Due to the very good N-doped graphene catalysts, many research groups have studied co-doping of N and another element (P, S, Fe) into graphene framework. The co-doping of nitrogen and another heteroatom to graphene structure, due to the synergistic effect between the heteroatoms, further promotes the catalytic activity [57, 58]. The investigation of such catalysts exhibits excellent yield, higher than 97%, for a variety of aromatic alcohols reactions. The large number of defects of graphene sheets in combination with the synergistic effect seems to be the key-idea not only for high catalytic activity, but for very good recyclability, as well.

The N-doped carbon atoms of graphene framework are positively charged, while the introduced N atoms are negatively charged. Then, the P-atoms that are introduced into the N-doped graphene framework increase the positive charge density activating more neighboring carbon atoms. This synergistic behavior between the N-doped carbon atoms and the P atoms activates the inert sp^2-hybridized carbon framework, enabling the transfer of H species during reactions [57]. The co-doped graphene material also develops hydrophilic nature and so the graphene-based organocatalysts can uniformly disperse into water.

Meanwhile, in some cases single P-doped graphene sheets have been reported as more effective organocatalysts [59]. The incorporation of P into graphene nanosheet at the optimum content can cause the reactants' adsorption at a low adsorption energy, causing also higher O-O bonds elongation, in the case of aerobic oxidative coupling of amines to imines (very significant immediates for pharmaceutical industry) [60, 61]. Moreover, the P-doped bilayer graphene compared to pristine graphene presents up to five times higher electron mobility [62, 63].

In organic synthesis, there are many reactions that form C-C bond that necessitate acid or basic catalysts. Thus, an N-doped graphene catalyst can have the role of the basic catalyst when we refer, for example, to the Michael and Henry additions [64]. The graphene framework delocalized π-orbitals can provide or receive electron density *via* the formed defects, which are found either at its edges or into the main vacancies of structure.

Zhang *et al.* [65] in their work highlight the contribution of GO organocatalyst, specifically to the Henry-Michael reactions, by comparing primary amine-GO performance with primary amine-activated carbon and primary amine-mesoporous silica organocatalysts. The special 2D open GO framework, which is accomplished with the inclusion of primary amines into GO basal planes, is of key importance for the reduction of the mass transfer resistance. This characteristic of GO special structure, in combination with its acid-base behavior, can show the GO-based heterogeneous organocatalysts higher performance (almost three times higher), compared to organocatalysts supported on other (than GO) material.

The silanization of the GO with organoalkoxysilane precursors, such as ureidopropyl (UDP) group, as an acid and the 3-[2--2-aminoethylamino)ethylamino]-propyl (AEP) group as a general base, have also been restrictively examined for the synthesis of bifunctional GO [66]. GO silanization has also been explored for SiGO synthesis and application into various applications, for example, in the biomedical field. This is a method with low cost that is used for the modification of surfaces that have a rich amount of hydroxyl groups, in order to obtain a covalent coating [67].

Zhang *et al.* [66] investigated the SiGO (GO-AEP-UDP, AEP:UDP=1:4) catalytic performance for the Henry reaction of 4-nitrobenzaldehyde and nitromethane to p,β-dinitrostyrene, reporting almost 90% products yield.

The active groups were homogeneously distributed on both GO plane sites. The combination of Si with amine groups can offer a bifunctional catalyst that is more reactive for reactions, such as CO_2 chemical fixation [68]. The hydrogen bonding donor along with the amine functional groups on the 2D GO surface increases the density of the active centers. The 2D dimensional amine rich GO surface having low resistance can expedite the epoxides and CO_2 diffusion and adsorption. Even after seven uses in series, the proposed hybrid organocatalyst retained its amine groups.

In order to further investigate the combination of Si with primary and tertiary amine groups onto the GO support, Zhang *et al.* [69] investigated a silylanization procedure of GO with the aid of amine-terminal silanes (Fig. **5**). The amine-terminal silanes were covalently bonded onto the GO surface. The as-fabricated heterogeneous catalysts were further examined for a trans-β-nitrostyrene forming reaction.

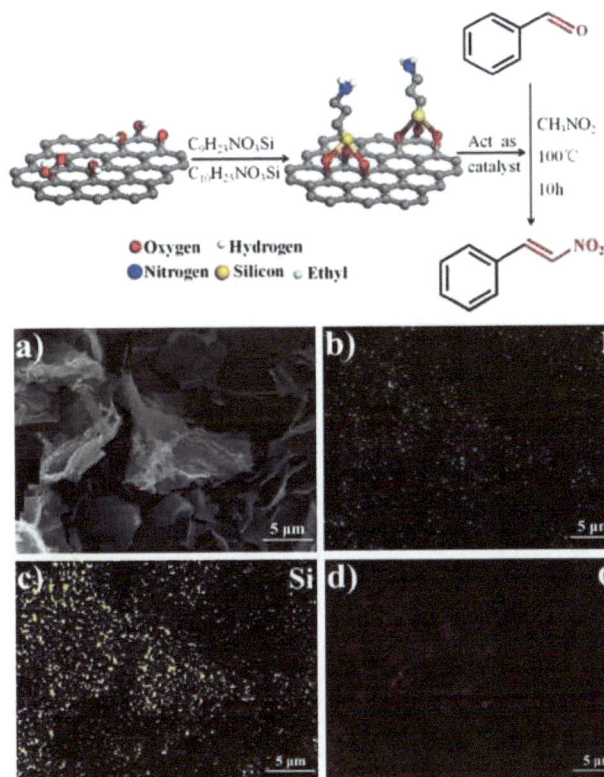

Fig. (5). Go functionalization with primary and tertiary amines (top) and SEM and EDS element analysis (bottom) [69].

The homogeneous distribution of the hydroxyl and epoxy groups onto GO surface along with the GO crumbling structure (Fig. **6B**) facilitated the access of reactants to the active sites. Additionally, the cooperation between the primary and tertiary amines, when added in 1:1 ratio, increased the hybrid catalysts selectivity. Specifically, it was observed that the primary amines activate the carbonyl compounds *via* the imine formation reactions, while the tertiary amines activate the nucleophiles, being indicated as a promising material for base catalytic reactions. However, in the last years the silanization process of GO does not seem to be the most preferable one, since other methods as above described, are more applicable.

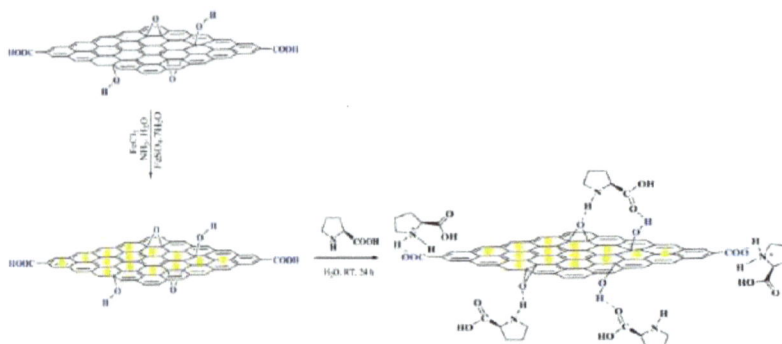

Fig. (6). Superparamagnetic graphene oxide/Fe$_3$O$_4$/l-proline [71].

MAGNETIC-BASED GRAPHENE OXIDE ORGANOCATALYSTS

The recycling and separation ease of a catalyst in industrial processes is of great importance for both environmental and economic reasons. The use of graphene oxide or reduced graphene oxide especially, as support to the organocatalysts, increased dramatically the product yields of many industrial reactions, also providing the possibility of recycling the as-synthesized catalysts.

Despite those offered advantages, there are some industrial processes that need special attention and so the separation of the catalyst from the reaction mixture should be easier and of very high accuracy. The key solution to such demanding solutions is the inclusion of magnetic nanoparticles in the functionalized GO sheet.

Recently Azarifar *et al.* [70] reported the novel synthesis of an Fe$_3$O$_4$-supported N-pyridin-4-amine-grafted graphene oxide. The inclusion of iron nanoparticles into the GO structure gave the final catalyst magnetic properties and so it could be easily removed from the solvent to a great extent. During the synthesis method initially, the GO functionalization procedure took place with the aminopyridine moieties following the common method of nucleophilic ring-opening of the epoxide groups; and then the iron oxide particles were added. The as-synthesized catalyst was evaluated successfully for the 4H-chromenes and dihydropyrano[2,3-c]pyrazole derivatives synthesis in water. It is characteristic that the inclusion of iron oxide particles did not modify the 2D GO structure sheets.

On the same philosophy, Keshavarz *et al.* [71] suggested a super magnetically non-covalently proline functionalized GO based catalyst for the synthesis of bis-pyrazole derivatives (Fig. **6**). The metal nanoparticles are uniformly distributed onto the functionalized support due to the aid of the functional groups onto the

GO support. According to this suggested synthesis method, the iron oxide nanoparticles were first anchored onto the GO support taking GO/Fe_3O_4 and then the pristine L-proline was mixed with GO/Fe_3O_4. The L-proline was non-covalently bonded due to the interaction between its secondary amine and carboxyl groups and the GO sheet hydroxyl, epoxy and carboxyl groups.

CONCLUSION

Heterogeneous organocatalysts catalytic activity and stereoselectivity when graphene oxide or reduced graphene oxide is adopted as support increase considerably. Moreover, the implementation of graphene material transformed the organic catalysts into recyclable materials that can be used many times, without a significant reduction to their activity. For those two reasons, it can be stated that graphene material brought the revolution to organic catalysis. Today, the tactic of organic molecule immobilization onto the graphene support is one of the main challenges that the research community must face. The other challenge is the finding of low-cost processes for catalyst removal. L-proline immobilization onto graphene oxide sheets *via* covalently bonding (hydrogen bonding or ions pair interactions) seems to be a promising method. However, the removal of proline-graphene catalysts is a complicated procedure and for this reason, investigation is made in order for magnetic nanoparticles to be introduced into its structure. Additionally, making use of graphene material, the research community synthesized acid-base bifunctional heterogeneous organocatalysts solving the problem with the co-presence of acid and base catalysts when it is necessary. Especially when eco-friendly substances such as piperazine are utilized, the bifunctional graphene catalyst becomes greener. Additionally, the poor solubility of graphene-based catalysts to solvents is also considered a great challenge that could be overcome by co-doping of graphene with non-metallic elements. The non-metals can develop graphene hydrophilic nature facilitating their uniform dispersion mostly into water. In general, graphene-based heterogeneous catalysts make the future very promising for organic catalysis field and for this reason, their properties should be further explored.

CONSENT FOR PUBLICATION

Not applicable.

CONFLICT OF INTEREST

The author declares no conflict of interest, financial or otherwise.

ACKNOWLEDGEMENTS

The author is grateful to the Greek Ministry of Education and the Laboratory of Alternative Energy Systems, Department of Mechanical Engineering, University of Thessaly, Volos, Greece.

REFERENCES

[1]   Chang, X.; Zhang, Q.; Guo, C. Asymmetric electrochemical transformations. *Angew. Chem. Int. Ed. Engl.,* **2020**, *59*(31), 12612-12622.
        [http://dx.doi.org/10.1002/anie.202000016] [PMID: 32057174]

[2]   Ghosh, M.; Shinde, V.S.; Rueping, M. A review of asymmetric synthetic organic electrochemistry and electrocatalysis: concepts, applications, recent developments and future directions. *Beilstein J. Org. Chem.,* **2019**, *15*(1), 2710-2746.
        [http://dx.doi.org/10.3762/bjoc.15.264] [PMID: 31807206]

[3]   Lin, Q.; Li, L.; Luo, S. Asymmetric electrochemical catalysis. *Chemistry,* **2019**, *25*(43), 10033-10044.
        [http://dx.doi.org/10.1002/chem.201901284] [PMID: 31026120]

[4]   Bolm, C.; Rantanen, T.; Schiffers, I.; Zani, L. Protonated chiral catalysts: versatile tools for asymmetric synthesis. *Angew. Chem. Int. Ed.,* **2005**, *44*(12), 1758-1763.
        [http://dx.doi.org/10.1002/anie.200500154] [PMID: 15754311]

[5]   Hirata, T.; Sato, I.; Yamashita, Y.; Kobayashi, S.J.C.C.; Asymmetric, C. (sp 3)–H functionalization of unactivated alkylarenes such as toluene enabled by chiral Brønsted base catalysts. *Commun. Chem.,* **2021**, *4*(1), 1-8.
        [http://dx.doi.org/10.1038/s42004-021-00459-5]

[6]   Frontana-Uribe, B.A.; Little, R.D.; Ibanez, J.G.; Palma, A.; Vasquez-Medrano, R.J.G.C. Organic electrosynthesis: a promising green methodology in organic chemistry. *Green Chem.,* **2010**, *12*(12), 2099-2119.
        [http://dx.doi.org/10.1039/c0gc00382d]

[7]   Wei, Z.; Chen, Y.; Wang, J.; Su, D.; Tang, M.; Mao, S.; Wang, Y. Cobalt Encapsulated in N-Doped Graphene Layers: An Efficient and Stable Catalyst for Hydrogenation of Quinoline Compounds. *ACS Catal.,* **2016**, *6*(9), 5816-5822.
        [http://dx.doi.org/10.1021/acscatal.6b01240]

[8]   Brouzgou, A. Oxygen evolution reaction. *Methods for Electrocatalysis*; Springer, **2020**, pp. 149-169.
        [http://dx.doi.org/10.1007/978-3-030-27161-9_6]

[9]   Tian, Y.; Liu, Q.; Liu, Y.; Zhao, R.; Li, G.; Xu, F. Catalyst-free Mannich-type reactions in water: Expedient synthesis of naphthol-substituted isoindolinones. *Tetrahedron Lett.,* **2018**, *59*(15), 1454-1457.
        [http://dx.doi.org/10.1016/j.tetlet.2018.02.083]

[10]   Lv, G.; Wang, H.; Yang, Y.; Deng, T.; Chen, C.; Zhu, Y.; Hou, X. Graphene Oxide: A Convenient Metal-Free Carbocatalyst for Facilitating Aerobic Oxidation of 5-Hydroxymethylfurfural into 2, 5-Diformylfuran. *ACS Catal.,* **2015**, *5*(9), 5636-5646.
        [http://dx.doi.org/10.1021/acscatal.5b01446]

[11]   Tan, R.; Li, C.; Luo, J.; Kong, Y.; Zheng, W.; Yin, D.J.J.c. An effective heterogeneous l-proline catalyst for the direct asymmetric aldol reaction using graphene oxide as support. *J. Catal.,* **2013**, *298*, 138-147.
        [http://dx.doi.org/10.1016/j.jcat.2012.11.024]

[12]   Shaikh, I.R. Organocatalysis: Key Trends in Green Synthetic Chemistry, Challenges, Scope towards Heterogenization, and Importance from Research and Industrial Point of View. *J. Catal.,* **2014**, *2014*, : 402860..

[13]   Zhang, W.; Li, Z.; Gu, H.; Li, Y.; Zhang, G.; Zhang, F.; Fan, X. l-proline covalently anchored on graphene oxide as an effective bifunctional catalyst for ketene forming reaction. *Chem. Eng. Sci.,* **2015**, *135*, 187-192.
[http://dx.doi.org/10.1016/j.ces.2015.04.050]

[14]   Sadiq, M.; Aman, R.; Saeed, K.; Ahmad, M. S.; Zia, M. A. J. M. R. i. C. Green and sustainable heterogeneous organo-catalyst for asymmetric aldol reactions. *Modern research in catalysis,* **2015**, *4*(02), 43.
[http://dx.doi.org/10.4236/mrc.2015.42006]

[15]   Zhou, M.; El-Sayed, E-S.M.; Ju, Z.; Wang, W.; Yuan, D.J.I.C.F. The synthesis and applications of chiral pyrrolidine functionalized metal–organic frameworks and covalent-organic frameworks. *Inorg. Chem. Front.,* **2020**, *7*(6), 1319-1333.
[http://dx.doi.org/10.1039/C9QI01103J]

[16]   Ding, S-Y.; Wang, W. Covalent organic frameworks (COFs): from design to applications. *Chem. Soc. Rev.,* **2013**, *42*(2), 548-568.
[http://dx.doi.org/10.1039/C2CS35072F] [PMID: 23060270]

[17]   Xu, H.; Gao, J.; Jiang, D. Stable, crystalline, porous, covalent organic frameworks as a platform for chiral organocatalysts. *Nat. Chem.,* **2015**, *7*(11), 905-912.
[http://dx.doi.org/10.1038/nchem.2352] [PMID: 26492011]

[18]   Shinde, D.B.; Kandambeth, S.; Pachfule, P.; Kumar, R.R.; Banerjee, R. Bifunctional covalent organic frameworks with two dimensional organocatalytic micropores. *Chem. Commun. (Camb.),* **2015**, *51*(2), 310-313.
[http://dx.doi.org/10.1039/C4CC07104B] [PMID: 25408225]

[19]   Zhang, H.; Lou, L.L.; Yu, K.; Liu, S.J.S. *Advances in Chiral Metal–Organic and Covalent Organic Frameworks for Asymmetric Catalysis*; Wiley Online Library, **2021**, p. 2005686.

[20]   Li, X. J. M. C. F. sp 2 carbon-conjugated covalent organic frameworks: synthesis, properties, and applications. *Matrerials Chemistry Frontiers,* **2021**.

[21]   Cote, A.P.; Benin, A.I.; Ockwig, N.W.; O'Keeffe, M.; Matzger, A.J.; Yaghi, O.M.J.s. *Porous, crystalline, covalent organic frameworks,* science.sciencemag.org **2005**.

[22]   Xu, H.; Chen, X.; Gao, J.; Lin, J.; Addicoat, M.; Irle, S.; Jiang, D. Catalytic covalent organic frameworks *via* pore surface engineering. *Chem. Commun. (Camb.),* **2014**, *50*(11), 1292-1294.
[http://dx.doi.org/10.1039/C3CC48813F] [PMID: 24352109]

[23]   Wang, J-C.; Kan, X.; Shang, J-Y.; Qiao, H.; Dong, Y-B.J.J.A.C.S. Catalytic Asymmetric Synthesis of Chiral Covalent Organic Frameworks from Prochiral Monomers for Heterogeneous Asymmetric Catalysis. *J. Am. Chem. Soc.,* **2020**, *142*(40), 16915-16920.
[http://dx.doi.org/10.1021/jacs.0c07461] [PMID: 32941016]

[24]   Zhi, Y.; Wang, Z.; Zhang, H.L.; Zhang, Q. Recent Progress in Metal-Free Covalent Organic Frameworks as Heterogeneous Catalysts. *Small,* **2020**, *16*(24), : e2001070..
[http://dx.doi.org/10.1002/smll.202001070] [PMID: 32419332]

[25]   Liu, Y.; Xuan, W.; Cui, Y. Engineering homochiral metal-organic frameworks for heterogeneous asymmetric catalysis and enantioselective separation. *Adv. Mater.,* **2010**, *22*(37), 4112-4135.
[http://dx.doi.org/10.1002/adma.201000197] [PMID: 20799372]

[26]   Bisbey, R.P.; Dichtel, W.R. Covalent organic frameworks as a platform for multidimensional polymerization. *ACS Cent. Sci.,* **2017**, *3*(6), 533-543.
[http://dx.doi.org/10.1021/acscentsci.7b00127] [PMID: 28691064]

[27]   Colson, J.W.; Mann, J.A.; DeBlase, C.R.; Dichtel, W.R. Patterned growth of oriented 2D covalent organic framework thin films on single-layer graphene. *J. Polym. Sci. A Polym. Chem.,* **2015**, *53*(2), 378-384.
[http://dx.doi.org/10.1002/pola.27399]

[28]   Colson, J. W.; Woll, A. R.; Mukherjee, A.; Levendorf, M. P.; Spitler, E. L.; Shields, V. B.; Spencer, M. G.; Park, J.; Dichtel, W. R. J. S. Oriented 2D covalent organic framework thin films on single-layer graphene. *Scinence.sciencemag* , **2011**, *332*(6062), 228-231.

[29]   Chen, Y.; Li, W.; Wang, X-H.; Gao, R-Z.; Tang, A-N.; Kong, D-M.J.M.C.F. Green synthesis of covalent organic frameworks based on reaction media. *Mater. Chem. Front.,* **2021**, *5*(3), 1253-1267. [http://dx.doi.org/10.1039/D0QM00801J]

[30]   Cui, D.; Perepichka, D.F.; MacLeod, J.M.; Rosei, F. Surface-confined single-layer covalent organic frameworks: design, synthesis and application. *Chem. Soc. Rev.,* **2020**, *49*(7), 2020-2038. [http://dx.doi.org/10.1039/C9CS00456D] [PMID: 32141466]

[31]   Xu, L.; Zhou, X.; Tian, W.Q.; Gao, T.; Zhang, Y.F.; Lei, S.; Liu, Z.F.J.A.C.I.E. Surface-confined single-layer covalent organic framework on single-layer graphene grown on copper foil. *Angew. Chem. Int. Ed. Engl.,* **2014**, *53*(36), 9564-9568. [http://dx.doi.org/10.1002/anie.201400273] [PMID: 25145927]

[32]   Hosseini, M-S.; Masteri-Farahani, M. Phenyl sulfonic acid functionalized graphene-based materials: Synthetic approaches and applications in organic reactions. *Tetrahedron Lett.,* **2021**, 132083. [http://dx.doi.org/10.1016/j.tet.2021.132083]

[33]   Maleki, A.; Hajizadeh, Z.; Abbasi, H.J.C.l. Surface modification of graphene oxide by citric acid and its application as a heterogeneous nanocatalyst in organic condensation reaction. *Carbon Letters,* **2018**, *27*, 42-49.

[34]   Wang, Y.; Li, H.; Wang, Y-Q.; Liu, Y.; Foxman, B.M.; Deng, L. Asymmetric Diels-Alder reactions of 2-pyrones with a bifunctional organic catalyst. *J. Am. Chem. Soc.,* **2007**, *129*(20), 6364-6365. [http://dx.doi.org/10.1021/ja070859h] [PMID: 17469829]

[35]   Li, Y.; Zhao, Q.; Ji, J.; Zhang, G.; Zhang, F.; Fan, X.J.R.A. Cooperative catalysis by acid–base bifunctional graphene. *RSC Advances,* **2013**, *3*(33), 13655-13658. [http://dx.doi.org/10.1039/c3ra41970c]

[36]   Khazaee, A.; Jahanshahi, R.; Sobhani, S.; Skibsted, J.; Sansano, J.M.J.G.C. Immobilized piperazine on the surface of graphene oxide as a heterogeneous bifunctional acid–base catalyst for the multicomponent synthesis of 2-amino-3-cyano-4 H-chromenes. *Green Chem.,* **2020**, *22*(14), 4604-4616. [http://dx.doi.org/10.1039/D0GC01274B]

[37]   Wang, D.; Lu, J.; Luo, L.; Jing, S.; Abbo, H.S.; Titinchi, S.J.J.; Chen, Z.; Tsiakaras, P.; Yin, S. Enhanced hydrogen evolution activity over microwave-assisted functionalized 3D structured graphene anchoring FeP nanoparticles. *Electrochim. Acta,* **2019**, *317*, 242-249. [http://dx.doi.org/10.1016/j.electacta.2019.05.153]

[38]   Rodrigo, E.; Alcubilla, B. G.; Sainz, R.; Fierro, J. G.; Ferritto, R.; Cid, M. B. J. C. C. Reduced graphene oxide supported piperazine in aminocatalysis. **2014**, *50*(47), 6270-6273. [http://dx.doi.org/10.1039/c4cc02701a]

[39]   Ramírez-Jiménez, R.; Franco, M.; Rodrigo, E.; Sainz, R.; Ferritto, R.; Lamsabhi, A.M.; Aceña, J.L.; Cid, M.B.J.J.M.C.A. Unexpected reactivity of graphene oxide with DBU and DMF. *J. Mater. Chem. A Mater. Energy Sustain.,* **2018**, *6*(26), 12637-12646. [http://dx.doi.org/10.1039/C8TA03529F]

[40]   Bourlinos, A.B.; Gournis, D.; Petridis, D.; Szabó, T.; Szeri, A.; Dékány, I.J.L. Graphite oxide: chemical reduction to graphite and surface modification with primary aliphatic amines and amino acids. *Langmuir,* **2003**, *19*(15), 6050-6055. [http://dx.doi.org/10.1021/la026525h]

[41]   Sobhani, S.; Zarifi, F.J.R.a. Pyridine-grafted graphene oxide: a reusable acid–base bifunctional catalyst for the one-pot synthesis of β-phosphonomalonates *via* a cascade Knoevenagel–phospha Michael addition reaction in water. *RSC Advances,* **2015**, *5*(117), 96532-96538.

[http://dx.doi.org/10.1039/C5RA13083B]

[42]   Becerril, H.A.; Mao, J.; Liu, Z.; Stoltenberg, R.M.; Bao, Z.; Chen, Y. Evaluation of solution-processed reduced graphene oxide films as transparent conductors. *ACS Nano,* **2008,** *2*(3), 463-470.
[http://dx.doi.org/10.1021/nn700375n] [PMID: 19206571]

[43]   Lim, J.; Yeo, H.; Kim, S.G.; Park, O-K.; Yu, J.; Hwang, J.Y.; Goh, M.; Ku, B-C.; Lee, H.S.; You, N-H. Pyridine-functionalized graphene/polyimide nanocomposites; mechanical, gas barrier, and catalytic effects. *Compos., Part B Eng.,* **2017,** *114,* 280-288.
[http://dx.doi.org/10.1016/j.compositesb.2016.12.057]

[44]   Choudhury, P.; Ghosh, P.; Basu, B. Amine-functionalized graphene oxide nanosheets (AFGONs): an efficient bifunctional catalyst for selective formation of 1,4-dihydropyridines, acridinediones and polyhydroquinolines. *Mol. Divers.,* **2020,** *24*(1), 283-294.
[http://dx.doi.org/10.1007/s11030-019-09949-0] [PMID: 30955149]

[45]   Brouzgou, A.; Gorbova, E.; Wang, Y.; Jing, S.; Seretis, A.; Liang, Z.; Tsiakaras, P. Nitrogen-doped 3D hierarchical ordered mesoporous carbon supported palladium electrocatalyst for the simultaneous detection of ascorbic acid, dopamine, and glucose. *Ionics,* **2019,** *25*(12), 6061-6070.
[http://dx.doi.org/10.1007/s11581-019-03116-z]

[46]   Wang, K.; Chen, H.; Zhang, X.; Tong, Y.; Song, S.; Tsiakaras, P.; Wang, Y. Iron oxide@graphitic carbon core-shell nanoparticles embedded in ordered mesoporous N-doped carbon matrix as an efficient cathode catalyst for PEMFC. *Appl. Catal. B,* **2020,** *264,* : 118468..
[http://dx.doi.org/10.1016/j.apcatb.2019.118468]

[47]   Li, B.; Li, Z.; Pang, Q.; Zhuang, Q.; Zhu, J.; Tsiakaras, P.; Shen, P.K. Synthesis and characterization of activated 3D graphene *via* catalytic growth and chemical activation for electrochemical energy storage in supercapacitors. *Electrochim. Acta,* **2019,** *324,* : 134878..
[http://dx.doi.org/10.1016/j.electacta.2019.134878]

[48]   Zhang, G.; Shi, Y.; Wang, H.; Jiang, L.; Yu, X.; Jing, S.; Xing, S.; Tsiakaras, P. A facile route to achieve ultrafine Fe2O3 nanorods anchored on graphene oxide for application in lithium-ion battery. *J. Power Sources,* **2019,** *416,* 118-124.
[http://dx.doi.org/10.1016/j.jpowsour.2019.01.091]

[49]   Liu, T.; Wang, K.; Song, S.; Brouzgou, A.; Tsiakaras, P.; Wang, Y. New Electro-Fenton Gas Diffusion Cathode based on Nitrogen-doped Graphene@Carbon Nanotube Composite Materials. *Electrochim. Acta,* **2016,** *194,* 228-238.
[http://dx.doi.org/10.1016/j.electacta.2015.12.185]

[50]   Gao, Y.; Tang, P.; Zhou, H.; Zhang, W.; Yang, H.; Yan, N.; Hu, G.; Mei, D.; Wang, J.; Ma, D. Graphene Oxide Catalyzed C-H Bond Activation: The Importance of Oxygen Functional Groups for Biaryl Construction. *Angew. Chem. Int. Ed. Engl.,* **2016,** *55*(9), 3124-3128.
[http://dx.doi.org/10.1002/anie.201510081] [PMID: 26809892]

[51]   Su, C.; Acik, M.; Takai, K.; Lu, J.; Hao, S.J.; Zheng, Y.; Wu, P.; Bao, Q.; Enoki, T.; Chabal, Y.J.; Loh, K.P. Probing the catalytic activity of porous graphene oxide and the origin of this behaviour. *Nat. Commun.,* **2012,** *3,* 1298.
[http://dx.doi.org/10.1038/ncomms2315] [PMID: 23250428]

[52]   Gao, Y.; Hu, G.; Zhong, J.; Shi, Z.; Zhu, Y.; Su, D. S.; Wang, J.; Bao, X.; Ma, D. J. A. C. I. E. Nitrogen-doped sp2-hybridized carbon as a superior catalyst for selective oxidation. **2013,** *52*(7), 2109-2113.
[http://dx.doi.org/10.1002/anie.201207918]

[53]   Kong, X-k.; Sun, Z-y.; Chen, M.; Chen, Q-J.E.; Science, E. Metal-free catalytic reduction of 4-nitrophenol to 4-aminophenol by N-doped graphene. *Energy Environ. Sci.,* **2013,** *6*(11), 3260-3266.
[http://dx.doi.org/10.1039/c3ee40918j]

[54]   Gao, Y.; Hu, G.; Zhong, J.; Shi, Z.; Zhu, Y.; Su, D.S.; Wang, J.; Bao, X.; Ma, D. Nitrogen-doped sp2-hybridized carbon as a superior catalyst for selective oxidation. *Angew. Chem. Int. Ed. Engl.,* **2013,**

52(7), 2109-2113.
[http://dx.doi.org/10.1002/anie.201207918] [PMID: 23307693]

[55]   Wang, Z.; Pu, Y.; Wang, D.; Wang, J-X.; Chen, J-F.J.F.C.S. Engineering, Recent advances on metal-free graphene-based catalysts for the production of industrial chemicals. *Front. Chem. Sci. Eng.,* **2018**, *12*(4), 855-866.
[http://dx.doi.org/10.1007/s11705-018-1722-y]

[56]   Shen, Q.; Chen, X.; Tan, Y.; Chen, J.; Chen, L.; Tan, S. interfaces, Metal-free N-formylation of amines with CO2 and Hydrosilane by nitrogen-doped graphene nanosheets. *ACS Appl. Mater. Interfaces,* **2019**, *11*(42), 38838-38848.
[http://dx.doi.org/10.1021/acsami.9b14509] [PMID: 31566364]

[57]   Yang, F.; Cao, Y.; Xu, C.; Xia, Y.; Chen, Z.; He, X.; Li, Y.; Yang, W.; Li, Y.J.A.J.O.C. Nitrogen and Phosphorus Co-Doped Graphene-Like Carbon Catalyzed Selective Oxidation of Alcohols. *Asian J. Org. Chem.,* **2019**, *8*(3), 422-427.
[http://dx.doi.org/10.1002/ajoc.201800677]

[58]   Xi, J.; Wang, Q.; Liu, J.; Huan, L.; He, Z.; Qiu, Y.; Zhang, J.; Tang, C.; Xiao, J.; Wang, S.J.J.C. N, P-dual-doped multilayer graphene as an efficient carbocatalyst for nitroarene reduction: a mechanistic study of metal-free catalysis. *J. Catal.,* **2018**, *359*, 233-241.
[http://dx.doi.org/10.1016/j.jcat.2018.01.003]

[59]   Yang, F.; Fan, X.; Wang, C.; Yang, W.; Hou, L.; Xu, X.; Feng, A.; Dong, S.; Chen, K.; Wang, Y.; Li, Y. P-doped nanomesh graphene with high-surface-area as an efficient metal-free catalyst for aerobic oxidative coupling of amines. *Carbon,* **2017**, *121*, 443-451.
[http://dx.doi.org/10.1016/j.carbon.2017.05.101]

[60]   Wang, X.; Li, Y.J.J.M.C.A. Nanoporous carbons derived from MOFs as metal-free catalysts for selective aerobic oxidations. *J. Mater. Chem. A Mater. Energy Sustain.,* **2016**, *4*(14), 5247-5257.
[http://dx.doi.org/10.1039/C6TA00324A]

[61]   Patel, M.A.; Luo, F.; Khoshi, M.R.; Rabie, E.; Zhang, Q.; Flach, C.R.; Mendelsohn, R.; Garfunkel, E.; Szostak, M.; He, H. P-Doped Porous Carbon as Metal Free Catalysts for Selective Aerobic Oxidation with an Unexpected Mechanism. *ACS Nano,* **2016**, *10*(2), 2305-2315.
[http://dx.doi.org/10.1021/acsnano.5b07054] [PMID: 26751165]

[62]   Wang, X.; Sun, G.; Routh, P.; Kim, D-H.; Huang, W.; Chen, P. Heteroatom-doped graphene materials: syntheses, properties and applications. *Chem. Soc. Rev.,* **2014**, *43*(20), 7067-7098.
[http://dx.doi.org/10.1039/C4CS00141A] [PMID: 24954470]

[63]   Maity, S.; Ram, F.; Dhar, B.B. Phosphorous-Doped Graphitic Material as a Solid Acid Catalyst for Microwave-Assisted Synthesis of β-Ketoenamines and Baeyer-Villiger Oxidation. *ACS Omega,* **2020**, *5*(26), 15962-15972.
[http://dx.doi.org/10.1021/acsomega.0c01231] [PMID: 32656417]

[64]   Candu, N.; Man, I.; Simion, A.; Cojocaru, B.; Coman, S.M.; Bucur, C.; Primo, A.; Garcia, H.; Parvulescu, V.I. Nitrogen-doped graphene as metal free basic catalyst for coupling reactions. *J. Catal.,* **2019**, *376*, 238-247.
[http://dx.doi.org/10.1016/j.jcat.2019.07.011]

[65]   Zhang, F.; Jiang, H.; Wu, X.; Mao, Z.; Li, H. Organoamine-functionalized graphene oxide as a bifunctional carbocatalyst with remarkable acceleration in a one-pot multistep reaction. *ACS Appl. Mater. Interfaces,* **2015**, *7*(3), 1669-1677.
[http://dx.doi.org/10.1021/am507221a] [PMID: 25556875]

[66]   Zhang, W.; Gu, H.; Li, Z.; Zhu, Y.; Li, Y.; Zhang, G.; Zhang, F.; Fan, X. J. J. o. M. C. A. General acid and base bifunctional graphene oxide for cooperative catalysis. *Chem. Soc. Rev.,* **2014**, *43*(20), 7067-7098.
[http://dx.doi.org/10.1039/C4TA01446D]

[67]   Vuppaladadium, S.S.R.; Agarwal, T.; Kulanthaivel, S.; Mohanty, B.; Barik, C.S.; Maiti, T.K.; Pal, S.;

Pal, K.; Banerjee, I. Silanization improves biocompatibility of graphene oxide. *Mater. Sci. Eng. C,* **2020**, *110*, 110647.
[http://dx.doi.org/10.1016/j.msec.2020.110647] [PMID: 32204077.]

[68]   Saptal, V.B.; Sasaki, T.; Harada, K.; Nishio-Hamane, D.; Bhanage, B.M.J.C. Hybrid amine-functionalized graphene oxide as a robust bifunctional catalyst for atmospheric pressure fixation of carbon dioxide using cyclic carbonates. *ChemSusChem,* **2016**, *9*(6), 644-650.
[http://dx.doi.org/10.1002/cssc.201501438] [PMID: 26840889]

[69]   Zhang, W.; Wang, S.; Ji, J.; Li, Y.; Zhang, G.; Zhang, F.; Fan, X. Primary and tertiary amines bifunctional graphene oxide for cooperative catalysis. *Nanoscale,* **2013**, *5*(13), 6030-6033.
[http://dx.doi.org/10.1039/c3nr01323e] [PMID: 23714770]

[70]   Azarifar, D.; Khaleghi-Abbasabadi, M.J.R.C.I. Fe 3 O 4-supported N-pyridin-4-amine-grafted graphene oxide as efficient and magnetically separable novel nanocatalyst for green synthesis of 4H-chromenes and dihydropyrano [2, 3-c] pyrazole derivatives in water. *Res. Chem. Intermed.,* **2019**, *45*(2), 199-222.
[http://dx.doi.org/10.1007/s11164-018-3597-4]

[71]   Keshavarz, M.; Zarei Ahmady, A.; Vaccaro, L.; Kardani, M. Non-Covalent Supported of l-Proline on Graphene Oxide/Fe_3O_4 Nanocomposite: A Novel, Highly Efficient and Superparamagnetically Separable Catalyst for the Synthesis of Bis-Pyrazole Derivatives. *Molecules,* **2018**, *23*(2), 330.
[http://dx.doi.org/10.3390/molecules23020330] [PMID: 29401720]

CHAPTER 3

Graphene Derived Materials as Catalysts for the Oxygen Reduction Reaction

Manisha Malviya[1,*], **Amisha Soni**[1] and **Sarvatej Kumar Maurya**[1]

[1] Department of Chemistry, Indian Institute of Technology, Banaras Hindu University, Varanasi, 221005, India

Abstract: The conventionally used electrocatalysts for oxygen reduction reaction (ORR) such as PGM and sulphide based, respectively, have contributed negligible or zero-emissions towards global warming. However, there is an urgent need to develop cost-effective, earth-abundant, efficient, non-poisonous and stable electrocatalysts. Through an exhaustive literature survey, it was observed that graphene derived materials are a promising candidate for ORR owing to exceptional electronic, physical and mechanical properties. Furthermore, the surface of graphene derived materials (GDM) can be modified to get desired physicochemical properties and induce electrocatalytic activity by adopting various synthesis methods insynthesis and post-synthesis by insertion of dopants, defect points, surface tuning, making composites, advanced wrapping structures. In the present chapter, fundamental ORR, graphene derived electrocatalysts, and the recent progress of ORR are discussed.

Keywords: Adsorption, Alkaline fuel cell, Binding energy, Defect sites, Direct methanol fuel cell, Doping, Electrocatalyst, Electrolytic media, Four-electron mechanism, Graphene, Graphene quantum dots, Metal-air batteries, MOF, Nanocomposites, Onset potential, Oxygen reduction reaction, Pyridine, Pyrrole, Spectroelectrochemical methods, Two-electron mechanism.

INTRODUCTION

The present century is a landmark from the technological perspective. Energy has become the most important topic for human societies and their sustainability. Worldwide, energy consumption is continuously rising to attain new heights of 50% in a quarter of a century [1]. Regular consumption of fossil fuels has increased greenhouse gas emissions like carbon dioxide due to incomplete or inefficient combustion of about 30 tonnes. Therefore, research to reform clean technology is required to transport, produce, and consume energy.

---

[*] **Corresponding author Manisha Malviya:** Department of Chemistry, Indian Institute of Technology, Banaras Hindu University, Varanasi, 221005, India; Email: manisha.apc@iitbhu.ac.in

To fulfil such innovation in fuel technology is not an easy endeavor. This technology, besides being practically feasible as per as performance view, efficiency must also be sustainable and affordable. Over the last few decades [2], graphene sheets were interesting for primary and applied research points of view due to their fascinating mechanical, electronic, and surface assets [3]. Graphene, a 2D monolayer of sp^2 hybridized carbon atoms, is arranged in a typical graphite-like hexagonal array also known as hexabenzocoronene (HBC-motif). Graphene became the center of attention as an electrocatalyst and find application in various essential research fields such as corrosion inhibitors and electrochemical sensors [4, 5, 5a, 5b, 5c]. Investigations have shown that pristine graphene has a low-performance toward O_2 reduction because of the saturated carbon atom. Graphene sheet has an extensive surface area, high charge carrier mobility, current, thermal conductivity, and heat conduction [6 - 7]. Different methods are known for the synthesis of graphene and graphene derived materials from graphite, such as *via* its mechanical cleavage [8], its chemical exfoliation [9], solvothermal method [10] and CVD [11]. It was found by experimental and theoretical studies that doping of graphene with p-type heteroatoms such as boron, nitrogen, and phosphorous can change its chemical reactivity and electronic property and generate new functions that can increase the electrocatalytic activity toward O_2 reduction [12 - 14]. The performance of graphene derived nanomaterials towards O_2 reduction depends on their well-defined morphology. The composite materials of graphene exhibit different properties from that pristine graphene due to oxygenated functional groups, defects, interlayer space, and aggregation of different graphene layers [15, 16]. Some commonly used terms put forth by Bianco *et al.* [17]:

a. Graphite oxide (oxidized form, avails oxo group on the basal planes and meant for interlayer spacing),
b. Graphene oxide (graphite oxide has been exfoliated to increase interlayer space further) and,
c. Reduced graphene (obtained by reduction of graphite oxide by electrochemical, thermal methods, *etc*).

Various derivatives of graphene have been used throughout the chapter as graphene derived materials.

GRAPHENE-BASED MATERIAL FOR O_2 REDUCTION

Heteroatom Doped Graphene Catalysts

Oxygen reduction reaction (ORR) is characterized by two-electron and four-electron pathways. It involves the reaction of surface adsorbed oxygen species with the metal ion or reactive surfaces, forming intermediate oxygen species like O-O*, *OOH, and *OH with appropriate overpotential, adsorption energy, binding energy and desorption energy. ORR is a sluggish reaction, and graphene shows poor reactivity for the two-electron pathway and is thereby unfavorable. Therefore, globally research is going on graphene-based materials doped with heteroatoms, mainly nitrogen has been studied vastly as a dopant atom [18]. There is a controversy among the scientific fraternity about the active site of doped graphene-based catalysts. The universal debate is about the integral role of structures and geometries of active species of metal/nonmetal likeable be the active site for reaction. It became questionable in acidic electrolytic media, where very few metal-free electrocatalysts have been synthesized for ORR. However, under alkaline electrolytic media where ORR is kinetically comparatively more favorable. Many metal-free electrocatalysts have been developed, which show at par electrocatalysis with platinum and its derivatives. It is a conventional view that in metal-free catalysts, especially N doped graphene, there is not only one, but structurally other active sites are present, and research has owed it to N doped carbon structures. It is essential to mention that this section focuses separately on graphene-based electrocatalysts and specifically does not discuss thermally treated nitrogen precursors and Fe/Co and (M-N-C electrocatalysts). Since these catalysts are helpful and play a crucial role under acidic electrolytic media, this chapter presents graphene-based electrocatalysts for alkaline media and finds application in alkaline fuel cells or metal-air batteries.

N- doped Graphene Derived Materials (NGDM)

Oxygen reduction reaction importantly follows steps viz., adsorption of molecular oxygen, breaking of O-O bond on to reactive sites and then desorption of intermediate oxygen species at NGDM. It is very well known to regulate these steps by governing different types of nitrogenous functional groups, and their number of atoms in the graphene derived materials [23 - 25]. In such structures, not only edges which are considered as low coordination sites, but also aromatic six-membered nitrogenous species, aromatic five-membered nitrogenous species, and extended graphene derived nitrogenous species were also reported as adsorption sites for molecular oxygen (Fig. 1a). The starting material and the method of synthesis of NGDM can regulate the adsorption and concentration of oxygen on the interactive surface of N doped GDM. However, the literature survey states that its synthetic methods have not been well regulated yet [23 - 26].

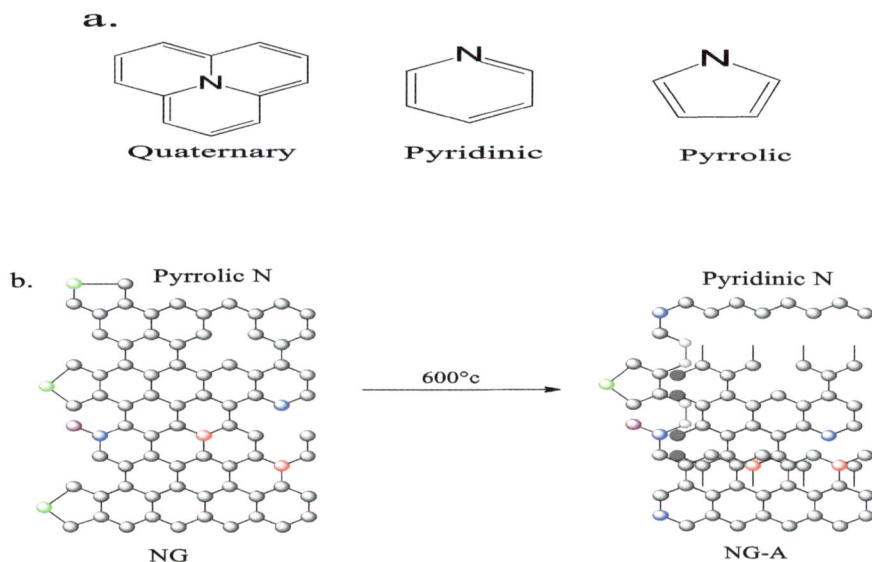

Fig. (1). (**a**) chemical structures of nitrogenous species (**b**) Heat treatment of aromatic five-membered N (pyrrole) active sites to aromatic six-membered N (pyridinic) active sites [27, 28].

Bai *et al.* [27] have reported work on oxygen reduction reaction and observed higher electrocatalytic activity of various N doped in acidic and alkaline electrolytic media. For this, the research group synthesized N doped and thermally treated NGDM as active materials using urea as a precursor for doping nitrogen into graphene oxide, followed by pyrolysis at various temperatures. It was observed that materials obtained after pyrolysis at 1000 °C were highly active towards ORR under acidic and basic electrolytic media. This may be due to the high defect ratio and incorporation of an optimized amount of N into graphene derived materials. Ma [28] reported oxygen reduction reaction under basic and acidic electrolytic conditions on NGDM using the solvothermal method and upgraded pyrrolic nitrogenous species to pyridinic one, as shown in Fig. (**1b**). The higher electrocatalytic activity was later observed due to more concentration of pyridinic nitrogen (about 31%) than pyrrolic one (about 16%). Rivera *et al.* [29] found experimental results supported by theoretical data for oxygen reduction reaction on nitrogen-containing graphene quantum dots (N-GQDs) and observed an increase in electron transfer and adsorption of different oxygen species onto edges. Besides their direct role in 4 electron transfer, N doped graphene was also employed as electrocatalyst support for many nanomaterials *viz.*, iron [30], cobalt [31], iron-cobalt bimetallic nanoparticles [32].

Moreover, nitrogen-doped graphene [31 - 33] showed higher methanol tolerance, good stability and electrocatalysis in acidic electrolytes. Furthermore, it is advocated that the collective effect of reactive metal and NGDM can enhance the electrocatalysis towards oxygen reduction reaction. Such electrocatalysts have been reported as composite materials in literature.

S- & S- and N- Co-Doped Graphene Derived Materials (SGDM/SNGDM)

S and C exhibit almost similar electronegativity at (Pauling) scale of 2.6 [2.58] and 2.6 [2.55], respectively, with a difference of an orbital between them. Both extrinsic and intrinsic defects can be observed with the Sulphur atom. It may affect the graphene structure by incorporating it into the structure as a foreign atom and may also perturb the crystalline order of the structure. Hence, SGDM creates defects that are proven preferential catalytic sites for oxygen reduction reaction [34, 35]. There is an urgent need to adapt and modify synthetic methodology to develop efficient electrocatalysts of S doped materials, which are also effective in acidic and alkaline electrolytic media [35]. Wang *et al.* [36] have used p-methyl benzenesulfonic acid as a reagent for sulfur onto magnesium oxide as a template to develop sulfur- graphene framework. Although the electrocatalytic activity of such framework was lower than the reported C supported Pt electrocatalyst, it exhibited a significant presence of sulphide bridges, *i.e.*, C-S-C, which were advocated as reactive sites for the oxygen reduction reaction. Moreover, it finds application in direct methanol fuel cell and has proven to be a suitable electrocatalyst for oxygen reduction reaction due to high methanol tolerance.

S and N co-doped graphene derived materials (NSGDM) are more effective than solo-doping of any of these towards oxygen reduction reaction. Based on the preference of graphene plane for each heteroatom, definite active site for oxygen reduction reaction is generated due to the incorporation of S and N atoms. These heteroatoms systematically modify the growth of reactive sites at the graphene framework [23, 26, 37]. Moreover, SNGDM followed the four-electron pathway as the primary operative mechanism towards oxygen reduction reaction in alkaline electrolyte [23, 29, 37]. Generally, cysteine has been used as precursors for S and N sources. It was observed that the incorporation of both heteroatoms has considerably reduced the onset potential for oxygen reduction reaction and yields high and constant limiting currents. Recently, it was found that nitrogen insertion into the graphene materials facilitates the adsorption and bond cleavage of O_2 due to the charge polarization of the N–C bond, which favors the four-electron pathway [38]. Also, the precise ratio of S and N (3:1) into the GDM increases the rate of the oxygen reduction reaction.

Nanocomposites of Graphene Derived Materials (ncGDM)

This part comprises composites of GDM with nanosized transition metals with N and or S dopants. Such ncGDM are versatile and can be employed for anodic and cathodic reaction in fuel cells [39, 40]. In order to achieve an increase in the performance of the oxygen reduction reaction, the properties of each species are combined to furnish a highly active hybrids framework. Fe-graphene nanocomposite materials are extensively studied as active electrocatalyst for the cathodic reaction in fuel cell due to their easy and cost-effective synthesis [30, 42]. Chen *et al*. [42] investigated Fe and N dual doping graphene nanoribbon composites for the oxygen reduction reaction. They found a high density of edge defect sites and a very stable framework. In this context, Niu *et al*. [30] prepared mesoporous Fe/N-doped graphene material with encapsulated Fe_3C nanoparticles. They showed high oxygen reduction reaction catalytic activity, long-time durability and high methanol tolerance in alkaline media. Graphene quantum dots (GQDs) are most important due to their high density of edge length, which are very active site material for different electrochemical reactions.

MOF and covalent organic frameworks are exciting species that may form 1, 2 or 3-dimensional porous structures materials and can be used as precursors to prepare graphene-based materials after a thermal treatment. An exciting example includes the preparation of sandwich-like structured N doped porous carbon graphene composite materials synthesized from sandwich-like structured zeolitic imidazolate material graphene oxides, which was occurred after a carbonization process at 900 °C. The extended graphene nano carbon revealed high surface area and porosity and higher catalytic performance than commercial carbon-supported Pt toward the oxygen reduction reaction in 0.1M KOH [43].

Karim Kakaei and Zahra Ostadi *et al*. [44] have electrodeposited Ni on carbon or graphene coated carbon using cyclic voltammetry and also Graphene on carbon electrode; the synthesis method for ORR explored for the first time. Thus obtained, this Ni onto graphene was found to have a remarkable active and stable catalyst for ORR in alkaline solution with electron transference of 3.91 observed by K-L equation. It was observed that such electrodeposition by CV leads to the formation of cauliflower structures of Ni nanoparticle on carbon paper and became shapeless on graphene surfaces on carbon paper with higher dispersion ranging from 60 and 40 nm.

Future of Graphene in ORR

Besides graphene there are other cousin's or siblings or derivatives of graphene which have found to exhibit tremendous success in ORR in alkaline media. These are: graphane, graphene-graphane, graphane-like structures [45], graphyne, graphdiyne, graphtriyne, graphtetrayne and more such derivatives. Graphane is one in which all carbon atoms are fully hydrogenated and sp^3 hybridized. Graphane has a formula of $(C_1H_1)_n$ in its ideal structure, without considering terminating hydrogen atoms. It was first time suggested in a theoretical investigation by Lu *et al.* [46], and the expected graphane structure was given by research group of Elias [47]. Graphane can be synthesized by exposing graphene to hydrogen plasma or a beam of hydrogen atoms for several hours. In this insulating graphane, the shared bond pair of electrons between hydrogen atoms and carbon atoms, pull the atoms out of the plane [48] and attribute many optoelectronic electric, magnetic properties. It is quite possible that along with graphane, partially hydrogenated graphene consisting of hydrogenated C atom in sp^3 hybridization and non-hydrogenated above and below the C atom in an alternating manner may form and it assumed to have a stable structure. Ao *et al.* [49] used the density functional theory to investigate the thermal stability of graphene/graphane nanoribbons, Lee and Grossman [50] have studied magnetic properties of graphene-graphane superlattices which were found to have zigzag interfaces and variable widths separately.

Another derivative is fluorographane $(C_1H_xF_{1-x-\delta})_n$, where symbol δ stands for the other elements most likely oxygen and traces of nitrogen, which may introduce during synthesis, is a new member of the graphene family that exhibits hydrophobicity and has tunable band gap by means of extent of fluorination [51]. Gusmao *et al.* [52] have synthesized flourographane by following a procedure of charging the graphane, evacuation, addition of a mixture of fluorine/nitrogen (20 vol% F_2) in a Monel autoclave with a Teflon liner. They have revealed that the carbon-to-fluoride ratio of fluorographane has a great impact on the ORR performance of the materials. The group has reported ORR on CHF (fluorographane) [24 h:5 bar], and CHF [24+ 24 h:5+12 bar] to have large potential shifts and much more intense currents. This also corresponds to electrocatalytic activity towards the ORR. With increasing fluoride content, the onset potential of ORR was observed to shift considerably less negative potentials in comparison to that of bare GC.

Baughman [53], was the first to theoretically predict Graphyne as a 2D carbon allotrope and composed of different types carbon atoms one with sp-hybridization and other with sp^2 -hybridization. Graphynes are 'expanded' forms of graphene, where C_2 units are inserted into $sp^2C–sp^2C$ bonds of the parent C_6 rings, through carbomerization, it may be complete C_2 insertion to give rise α-graphyne or incomplete C_2 insertion to give γ-graphyne, respectively. Further advanced graphyne materials are graphdiyne which have attracted great attention based on their unique physical, chemical, and mechanical properties due to their extensive pores and intercalation with other metals and non-metals. Due to presence of pores, these possess larger specific surface area and pores allowing free penetrations of small molecules to significantly improve mass-transportation. Extensive studies have reported that graphyne materials can efficiently facilitate the ORR process [54, 55].

Graphdiyne (GDY) contains specifically three- different types of hybridised C-C bonds *viz.* C (sp²) -C (sp²) bond on the aromatic benzene ring, other is C=C and C≡C bonds, and the most important two adjacent C (sp) -C (sp) bond. The acetylenic bond formed by hybridization of sp and sp² of benzene ring constitute graphdiyne molecule with the 2D planar structure of a single atomic layer. Graphdiyne molecules are stacked through weak vander wall forces of attraction and π-π interaction to form a layered structure. It has a large triangular shaped ring of eighteen carbon atoms which provides a three-dimensional channel and porous structure in the layered molecule. The planar sp² and sp hybrid configuration endows a uniformly dispersed pore configuration, a tunable electronic structure property and high degree of π conjugation. Therefore, strictly speaking, graphdiyne not only has characteristics of 2D planar materials similar to graphene, but also the characteristic of 3D porous materials. Such a rigid planar structure, uniform pore and other unique electronic properties make graphdiyne very suitable for the energy catalysts, storage and separation. Many research groups have studied graphdiyne towards ORR by theoretical and experimental investigations and observed that graphdiyne do not possess uniform charge distributions at each carbon atom which shows the existence of positively charged carbon atoms or active sites which are favorable to the adsorption of O_2 and OOH^+ species for ORR. The intrinsic porous structure of graphdiyne also contributes to the O_2 adsorption and thus facilitate the ORR. Gao *et al.* [56], firstly made theoretical predictions using DFT calculations, and then validated experimentally using Fe single atom catalyst with graphdiyne and achieved a high catalytic activity towards ORR similar to or even better than commercial Pt/C (20 wt%). The cathodic peak potential of Fe/ graphdiyne was observed at lower peak potential as compare to Pt/C with higher cathodic peak current. Doping of graphdiyne materials with heteroatoms found to improve catalytic activity for example Zhang's group [57] have synthesized N-doped graphdiyne which exhibited better stability and increased tolerance to the cross-over effect. They have also investigated different heteroatoms like S, B, F and corresponding codoped with Nitrogen [58 - 60]. Another research group of Zhang [61], have demonstrated theoretically and experimentally that the sp-N atom is the most beneficial active site for oxygen adsorption and showed the highest ORR performance in alkaline solution among different N types of metal-free carbon electrocatalysts.

Wang *et al.* [62] have reported Co and N codoped graphdiyne materials where the high Co content signifies the strong bonding effect between graphdiyne and Co. It was found that in 1% Co-N-GDY composition, at one hand existence of Co atoms has efficiently improved the electrocatalytic performance; on the other hand, the N-doping modified the electron configuration of graphdiyne and owed to efficient catalytic activity for ORR in 0.1 M KOH alkaline media.

Various DFT studies focused on oxygen reduction reaction mechanism were carried out on *N*-doped HsGDY (N–HsGDY) obtained from hydrogen substituted graphdiyne (HsGDY) [63] and the site-specific N doping of graphdiyne including grap-N, sp-N (I) and sp-N (II)-doped graphdiyne were systematically investigated as ORR electrocatalysts [64]. The study revealed that the N- doping improved the charge redistribution is and thus has a significant effect on the onset potential of ORR.

Though all the cousins of graphene are theoretically and experimentally proven as potent electrocatalysts for ORR in alkaline media, there are many challenges like (i) large-scale preparation of highly crystalline single- and multi-layer graphdiyne; (ii) understanding the reaction kinetics, process, and mechanism of graphdiyne growth and reveal the key growth factors of graphdiyne from amorphous to crystalline, (iii) Interfacial analysis for characterizing the interface structure and interfacial synergetic effects of various sp-sp, sp^2-sp and sp^2-sp^2 with intercalated and doped structures can be understood.

CONCLUSIONS AND OUTLOOK

O_2 reduction is the most critical research field because of the commercialization of fuel cells. In this regard, graphene derived materials have emerged as one of the most promising materials for oxygen reduction reactions. Its exceptional properties and versatility are employed to prepare electrocatalysts with high catalytic activity toward the oxygen reduction reaction.

The author has first discoursed the fundamentals of O_2 reduction, followed by a survey on the recent progress in graphene-based catalysts for O_2 reduction. The chapter presents literature for synthesis and electro catalytic activity of NGDM, SGDM, SNGDM, ncGDM, MOF and COF intercalated graphene derived materials towards the oxygen reduction reaction. Furthermore, different strategies used to prepare the derivatives of graphene-based materials for O_2 were also discussed. Pristine graphene is nonreactive, however, its derivatives like graphane, graphyne, graphdiyne, graphtriyne and graphtetrayne are semiconducting to conducting materials found to play an important role in the development of GDM, which was illustrated by theoretical and experimental studies. Such graphene derived materials possess a tunable interface and pores for oxygen adsorption and binding for ORR. The interaction between the surface-bound oxygen intermediates can be optimized on such future graphene structures. Therefore, it can be generalized that graphene-based electrocatalysts can be successfully employed to tune or modify the interface for the oxygen reduction reaction. In addition, the *in-situ* and *ex-situ* spectroelectrochemical and microscopic methods can provide a deep understanding of the ORR mechanism. Studying the physical and mechanical properties of graphene-based catalysts would be a new field of research in the application of O_2 reduction. It may play a significant role in decarbonization and also in the commercialization of fuel cells.

CONSENT FOR PUBLICATION

Not applicable.

CONFLICT OF INTEREST

The author declares no conflict of interest, financial or otherwise.

ACKNOWLEDGEMENTS

The author acknowledges Mr. Satyendra Pratap Singh for fruitful discussion.

REFERENCES

[1]   Bi, E.; Chen, H.; Yang, X.; Peng, W.; Grätzel, M.; Han, L. A quasi core-shell nitrogen-doped graphene/cobalt sulfide conductive catalyst for highly efficient dye-sensitized solar cells. *Energy Environ. Sci.,* **2014**, *7,* 2637.
        [http://dx.doi.org/10.1039/C4EE01339E]

[2]   Novoselov, K.S.; Geim, A.K.S.V.; Morozov, S.V.; Jiang, D.; Zhang, Y.; Dubonos, S.V.; Grigorieva, I.V.; Firsov, A.A. Electric field effect in atomically thin carbon films. *Science,* **2004**, *306*(5696), 666-669.
        [http://dx.doi.org/10.1126/science.1102896] [PMID: 15499015]

[3]   Geim, A.K.; Novoselov, K.S. The rise of graphene. *Nat. Mater.,* **2007**, *6*(3), 183-191.
        [http://dx.doi.org/10.1038/nmat1849] [PMID: 17330084]

[4]   Zhang, Y.; Tan, Y-W.; Stormer, H.L.; Kim, P. Experimental observation of the quantum Hall effect and Berry's phase in graphene. *Nature,* **2005**, *438*(7065), 201-204.
        [http://dx.doi.org/10.1038/nature04235] [PMID: 16281031]

[5]   Berger, C.; Song, Z.; Li, X.; Wu, X.; Brown, N.; Naud, C.; Mayou, D.; Li, T.; Hass, J.; Marchenkov, A.N.; Conrad, E.H.; First, P.N.; de Heer, W.A. Electronic confinement and coherence in patterned epitaxial graphene. *Science,* **2006**, *312*(5777), 1191-1196.
        [http://dx.doi.org/10.1126/science.1125925] [PMID: 16614173]

[5a]   Quraishi, M.A.; Gupta, R.K.; Malviya, M.; Verma, C.; Gupta, N.K. Pyridine based functionalized graphene oxides as new corrosion inhibitors for mild steel: Experimental and DFT approach. *RSC Advances,* **2017**, *7,* 39063.
        [http://dx.doi.org/10.1039/C7RA05825J]

[5b]   Gupta, R.K.; Malviya, M.; Ansari, K.R.; Lgaz, H.; Chauhan, D.S. Functionalized graphene oxide as a new generation corrosion inhibitor for industrial pickling process: DFT and experimental approach. *Mater. Chem. Phys.,* **2019**, *236,* 121727.
        [http://dx.doi.org/10.1016/j.matchemphys.2019.121727]

[5c]   Hasan, S.H.; Malviya, M. Vijay Kumar, Gupta, R.K.; Gundampati, R.K.; Devendra. Enhanced electron transfer mediated detection of hydrogen peroxide using a silver nanoparticle–reduced graphene oxide– polyaniline fabricated electrochemical sensor. *RSC Advances,* **2017**, *8*, 619-663.

[6]   Ghosh, S.; Calizo, I.; Teweldebrhan, D. Extremely high thermal conductivity of graphene: prospects for thermal management applications in nanoelectronic circuits. *Appl. Phys. Lett.,* **2008**, *92*, 151911. [http://dx.doi.org/10.1063/1.2907977]

[7]   Stoller, M.D.; Park, S.; Zhu, Y.; An, J.; Ruoff, R.S. Graphene-based ultracapacitors. *Nano Lett.,* **2008**, *8*(10), 3498-3502. [http://dx.doi.org/10.1021/nl802558y] [PMID: 18788793]

[8]   Novoselov, K.S.; Geim, A.K.; Morozov, S.V.; Jiang, D.; Zhang, Y.; Dubonos, S.V.; Grigorieva, I.V.; Firsov, A.A. Electric field effect in atomically thin carbon films. *Science,* **2004**, *306*(5696), 666-669. [http://dx.doi.org/10.1126/science.1102896] [PMID: 15499015]

[9]   Tung, V.C.; Allen, M.J.; Yang, Y.; Kaner, R.B. High-throughput solution processing of large-scale graphene. *Nat. Nanotechnol.,* **2009**, *4*(1), 25-29. [http://dx.doi.org/10.1038/nnano.2008.329] [PMID: 19119278]

[10]   Choucair, M.; Thordarson, P.; Stride, J.A. Gram- scale graphene production based on solvothermal synthesis and sonication. *Nat. Nanotechnol.,* **2009**, *4*, 30-33. [http://dx.doi.org/10.1038/nnano.2008.365] [PMID: 19119279]

[11]   Kim, K.S.; Zhao, Y.; Jang, H.; Lee, S.Y.; Kim, J.M.; Kim, K.S.; Ahn, J.H.; Kim, P.; Choi, J.Y.; Hong, B.H. Large-scale pattern growth of graphene films for stretchable transparent electrodes. *Nature,* **2009**, *457*(7230), 706-710. [http://dx.doi.org/10.1038/nature07719] [PMID: 19145232]

[12]   Wu, T.; Lin, J.; Cheng, Y.; Tian, J.; Wang, S.; Xie, S.; Pei, Y.; Yan, S.; Qiao, M.; Xu, H.; Zong, B. Porous Graphene-Confined Fe-K as Highly Efficient Catalyst for CO_2 Direct Hydrogenation to Light Olefins. *ACS Appl. Mater. Interfaces,* **2018**, *10*(28), 23439-23443. [http://dx.doi.org/10.1021/acsami.8b05411] [PMID: 29956535]

[13]   Ning, H.; Mao, Q.; Wang, W. N-doped reduced graphene oxide supported Cu_2O nanocubes as high active catalyst for CO_2 electroreduction to C_2H_4. *J. Alloys Compd.,* **2019**, *785*, 7-12. [http://dx.doi.org/10.1016/j.jallcom.2019.01.142]

[14]   Tao, H.; Sun, X.; Back, S.; Han, Z.; Zhu, Q.; Robertson, A.W.; Ma, T.; Fan, Q.; Han, B.; Jung, Y.; Sun, Z. Doping palladium with tellurium for the highly selective electrocatalytic reduction of aqueous CO_2 to CO. *Chem. Sci. (Camb.),* **2017**, *9*(2), 483-487. [http://dx.doi.org/10.1039/C7SC03018E] [PMID: 29629117]

[15]   Zhang, N.; Yang, M-Q.; Liu, S.; Sun, Y.; Xu, Y-J. Waltzing with the versatile platform of graphene to synthesize composite photocatalysts. *Chem. Rev.,* **2015**, *115*(18), 10307-10377. [http://dx.doi.org/10.1021/acs.chemrev.5b00267] [PMID: 26395240]

[16]   Yang, M-Q.; Zhang, N.; Pagliaro, M.; Xu, Y-J. Artificial photosynthesis over graphene-semiconductor composites. Are we getting better? *Chem. Soc. Rev.,* **2014**, *43*(24), 8240-8254. [http://dx.doi.org/10.1039/C4CS00213J] [PMID: 25200332]

[17]   Bianco, A.; Cheng, H-M.; Enoki, T.; Gogotsi, Y.; Hurt, R.H.; Koratkar, N.; Kyotani, T.; Monthioux, M.; Park, C.R.; Tascon, J.M.D.; Zhang, J. All in the graphene family – A recommended nomenclature for two-dimensional carbon materials. *Carbon,* **2013**, *65*, 1-6. [http://dx.doi.org/10.1016/j.carbon.2013.08.038]

[18]   Wang, H.; Maiyalagan, T.; Wang, X. Review on Recent Progress in Nitrogen-Doped Graphene: Synthesis, Characterization, and Its Potential Applications. *Catalysis,* **2012**, *2*, 781-794. [http://dx.doi.org/10.1021/cs200652y]

[19]   Hummers, W.S.; Offeman, R.E. Preparation of Graphitic Oxide. *J. Am. Chem. Soc.,* **1958**, *80*, 1339-1339.

[http://dx.doi.org/10.1021/ja01539a017]

[20] Wang, L.; Ambrosi, A.; Pumera, M. "Metal-free" catalytic oxygen reduction reaction on heteroatom-doped graphene is caused by trace metal impurities. *Angew. Chem. Int. Ed. Engl.,* **2013**, *52*(51), 13818-13821.
[http://dx.doi.org/10.1002/anie.201309171] [PMID: 24490277]

[21] Masa, J.; Zhao, A.; Xia, W.; Sun, Z.; Mei, B.; Muhler, M.; Schuhmann, W. Trace metal residues promote the activity of supposedly metal-free nitrogen-modified carbon catalysts for the oxygen reduction reaction. *Electrochem. Commun.,* **2013**, *34*, 113-116.
[http://dx.doi.org/10.1016/j.elecom.2013.05.032]

[22] Masa, J.; Zhao, A.; Xia, W.; Sun, Z.; Mei, B.; Muhler, M. Schuhmann. Metal-free catalysts for oxygenreduction in alkaline electrolytes: Influence of the presence of Co, Fe, Mn and Ni inclusions. *Electrochim. Acta,* **2014**, *128*, 271-278.
[http://dx.doi.org/10.1016/j.electacta.2013.11.026]

[23] Pan, F.; Duan, Y.; Zhang, X.; Zhang, J. A facile synthesis of nitrogen/sulfur co doped graphene for the oxygen reduction reaction. *ChemCatChem,* **2016**, *8*, 163-170.
[http://dx.doi.org/10.1002/cctc.201500893]

[24] Yadegari, A.; Samiee, L.; Tasharrofi, S. Tajik, S.; Rashidi, A.; Shoghi, F.; Rasoulianboroujeni, M.; Tahriri, M.; Rowley- Neale, S.J.; Banks, C.E. Nitrogen doped nanoporous graphene: an efficient metal-free electrocatalyst for the oxygen reduction reaction. *RSC Advances,* **2017**, *7*, 55555-55566.
[http://dx.doi.org/10.1039/C7RA10626B]

[25] Zhang, B.; Xiao, C.; Xiang, Y.; Dong, B.; Ding, S.; Tang, Y. Nitrogen-doped graphene quantum dots anchored on thermally reduced graphene oxide as an electrocatalyst for the oxygen reduction reaction. *ChemElectroChem,* **2016**, *3*, 864-870.
[http://dx.doi.org/10.1002/celc.201600123]

[26] Perivoliotis, D.K.; Tagmatarchis, N. Recent advancements in metal-based hybrid electrocatalysts supported on graphene and related 2D materials for the oxygen reduction reaction. *Carbon,* **2017**, *118*, 493-510.
[http://dx.doi.org/10.1016/j.carbon.2017.03.073]

[27] Bai, X.; Shi, Y.; Guo, J.; Gao, L.; Wang, K.; Du, Y.; Ma, T. Catalytic activities enhanced by abundant structural defects and balanced N distribution of N-doped graphene in oxygen reduction reaction. *J. Power Sources,* **2016**, *306*, 85-91.
[http://dx.doi.org/10.1016/j.jpowsour.2015.10.081]

[28] Ma, R.; Ren, X.; Xia, B.Y.; Zhou, Y.; Sun, C.; Liu, Q.; Liu, J.; Wang, J. Novel synthesis of N-doped graphene as an efficient electrocatalyst towards oxygen reduction. *Nano Res.,* **2016**, *9*, 808-819.
[http://dx.doi.org/10.1007/s12274-015-0960-2]

[29] Zhang, P.; Hu, Q.; Yang, X.; Hou, X.; Mi, J.; Liu, L.; Dong, M. Size effect of oxygen reduction reaction on nitrogen-doped graphene quantum dots. *RSC Advances,* **2018**, *8*, 531-536.
[http://dx.doi.org/10.1039/C7RA10104J]

[30] Niu, Y.; Huang, X.; Hu, W. Fe_3C nanoparticle decorated Fe/N doped graphene for efficient oxygen reduction reaction electrocatalysis. *J. Power Sources,* **2016**, *332*, 305-311.
[http://dx.doi.org/10.1016/j.jpowsour.2016.09.130]

[31] Fu, X.; Choi, J-Y.; Zamani, P.; Jiang, G.; Hoque, M.A.; Hassan, F.M.; Chen, Z. Co– N decorated hierarchically porous graphene aerogel for efficient oxygen reduction reaction in acid. *ACS Appl. Mater. Interfaces,* **2016**, *8*(10), 6488-6495.
[http://dx.doi.org/10.1021/acsami.5b12746] [PMID: 26937737]

[32] Palaniselvam, T.; Kashyap, V.; Bhange, S.N.; Baek, J-B.; Kurungot, S. Nanoporous graphene enriched with Fe/Co-N active sites as a promising oxygen reduction electrocatalyst for anion exchange membrane fuel cells. *Adv. Funct. Mater.,* **2016**, *26*, 2150-2162.
[http://dx.doi.org/10.1002/adfm.201504765]

[33] Zhang, G.; Lu, W.; Cao, F.; Xiao, Z.; Zheng, X. N-doped graphene coupled with Co nanoparticles as an efficient electrocatalyst for oxygen reduction in alkaline media. *J. Power Sources,* **2016**, *302*, 114-125.
[http://dx.doi.org/10.1016/j.jpowsour.2015.10.055]

[34] Poh, H.L.; Šimek, P.; Sofer, Z.; Pumera, M. Sulfur-doped graphene *via* thermal exfoliation of graphite oxide in H_2S, SO_2, or CS_2 gas. *ACS Nano,* **2013**, *7*(6), 5262-5272.
[http://dx.doi.org/10.1021/nn401296b] [PMID: 23656223]

[35] Klingele, M.; Pham, C.; Vuyyuru, K.R.; Britton, B.; Holdcroft, S.; Fischer, A.; Thiele, S. Sulfur doped reduced graphene oxide as metal-free catalyst for the oxygen reduction reaction in anion and proton exchange fuel cells. *Electrochem. Commun.,* **2017**, *77*, 71-75.
[http://dx.doi.org/10.1016/j.elecom.2017.02.015]

[36] Wang, C.; Yang, F.; Xu, C.; Cao, Y.; Zhong, H.; Li, Y. Sulfur-doped porous graphene frameworks as an efficient metal-free electrocatalyst for oxygen reduction reaction. *Mater. Lett.,* **2018**, *214*, 209-212.
[http://dx.doi.org/10.1016/j.matlet.2017.11.120]

[37] Wu, M.; Dou, Z.; Chang, J.; Cui, L. Nitrogen and sulfur co-doped graphene aerogels as an efficient metal-free catalyst for oxygen reduction reaction in an alkaline solution. *RSC Advances,* **2016**, *6*, 22781-22790.
[http://dx.doi.org/10.1039/C5RA22136F]

[38] Rivera, M.L.; Fajardo, S.; Arévalo, D.M.; García, G.; Pastor, E. S- and N-doped graphene nanomaterials for the oxygen reduction reaction. *Catalysts,* **2017**, *7*, 278-289.
[http://dx.doi.org/10.3390/catal7090278]

[39] Sun, M.; Liu, H.; Liu, Y.; Qu, J.; Li, J. Graphene-based transition metal oxide nanocomposites for the oxygen reduction reaction. *Nanoscale,* **2015**, *7*(4), 1250-1269.
[http://dx.doi.org/10.1039/C4NR05838K] [PMID: 25502117]

[40] Ratso, S.; Kruusenberg, I.; Joost, U.; Saar, R.; Tammeveski, K. Enhanced oxygen reduction reaction activity of nitrogen-doped graphene/multi-walled carbon nanotube catalysts in alkaline media. *Int. J. Hydrogen Energy,* **2016**, *41*, 22510-22519.
[http://dx.doi.org/10.1016/j.ijhydene.2016.02.021]

[41] Ratso, S.; Kruusenberg, I.; Vikkisk, M.; Joost, U.; Shulga, E.; Kink, I.; Kallio, T.; Tammeveski, K. Highly active nitrogen-doped few- layer graphene/carbon nanotube composite electrocatalyst for oxygen reduction reaction in alkaline media. *Carbon,* **2014**, *73*, 361-370.
[http://dx.doi.org/10.1016/j.carbon.2014.02.076]

[42] Chen, C.; Zhang, X.; Zhou, Z-Y.; Yang, X-D.; Zhang, X-S.; Sun, S-G. Highly active Fe, N co- doped graphene nanoribbon/carbon nanotube composite catalyst for oxygen reduction reaction. *Electrochim. Acta,* **2016**, *222*, 1922-1930.
[http://dx.doi.org/10.1016/j.electacta.2016.12.005]

[43] Liu, S.; Zhang, H.; Zhao, Q.; Zhang, X.; Liu, R.; Ge, X.; Wang, G.; Zhao, H.; Cai, W. Metal-organic framework derived nitrogen-doped porous carbon@graphene sandwich-like structured composites as bifunctional electrocatalysts for oxygen reduction and evolution reactions. *Carbon,* **2016**, *106*, 74-83.
[http://dx.doi.org/10.1016/j.carbon.2016.05.021]

[44] Kakaei, K.; Ostadi, Z. Nickel nanoparticles coated on the exfoliated graphene layer as an efficient and stable catalyst for oxygen reduction and hydrogen evolution in alkaline media. *Mater. Res. Express,* **2020**, *7*, 055504.
[http://dx.doi.org/10.1088/2053-1591/ab8c70]

[45] Zhou, C.; Chen, S.; Lou, J.; Wang, J.; Yang, Q.; Liu, C.; Huang, D.; Zhu, T. Graphene's cousin: the present and future of graphane. *Nanoscale Res. Lett.,* **2014**, *9*(1), 26.
[http://dx.doi.org/10.1186/1556-276X-9-26] [PMID: 24417937]

[46] Lu, N.; Li, Z.; Yang, J. Electronic structure engineering *via* on-plane chemical functionalization: a

comparison study on two-dimensional polysilane and graphane. *PhysChemComm,* **2009**, *113*, 16741.

[47]   Elias, D.C.; Nair, R.R.; Mohiuddin, T.M.G.; Morozov, S.V.; Blake, P.; Halsall, M.P.; Ferrari, A.C.; Boukhvalov, D.W.; Katsnelson, M.I.; Geim, A.K.; Novoselov, K.S. Control of graphene's properties by reversible hydrogenation: evidence for graphane. *Science,* **2009**, *323*(5914), 610-613.
       [http://dx.doi.org/10.1126/science.1167130] [PMID: 19179524]

[48]   Savchenko, A. Materials science. Transforming graphene. *Science,* **2009**, *323*(5914), 589-590.
       [http://dx.doi.org/10.1126/science.1169246] [PMID: 19179516]

[49]   Ao, Z.M.; Hernández-Nieves, A.D.; Peeters, F.M.; Li, S. Enhanced stability of hydrogen atoms at the graphene/graphane interface of nanoribbons. *Appl. Phys. Lett.,* **2010**, *97*, 233109.
       [http://dx.doi.org/10.1063/1.3525377]

[50]   Lee, J-H.; Grossman, J.C. Magnetic properties in graphene-graphane superlattices. *Appl. Phys. Lett.,* **2010**, *97*, 97.

[51]   Sofer, Z.; Simek, P.; Maznek, V. Sembera. F.; Janousek, Z. Pumera. M. Fluorographane ($C_1H_xF_{1-x-\delta}$)$_{(n)}$: synthesis and properties. *Chem. Commun. (Camb.),* **2015**, *51*, 5633-5636.
       [http://dx.doi.org/10.1039/C4CC08844A] [PMID: 25693806]

[52]   Gusmão, R.; Sofer, Z.; Šembera, F.; Janoušek, Z.; Pumera, M. Electrochemical Fluorographane: Hybrid Electrocatalysis of Biomarkers, Hydrogen Evolution, and Oxygen Reduction. *Chemistry,* **2015**, *21*(46), 16474-16478.
       [http://dx.doi.org/10.1002/chem.201502535] [PMID: 26442653]

[53]   Baughman, R.H.; Eckhardt, H.; Kertesz, M. Structure-property predictions for new planar forms of carbon: Layered phases containing sp2 and sp atoms. *J. Chem. Phys.,* **1987**, *87*(11), 6687-6699.
       [http://dx.doi.org/10.1063/1.453405]

[54]   Liu, R.; Liu, H.; Li, Y.; Yi, Y.; Shang, X.; Zhang, S.; Yu, X.; Zhang, S.; Cao, H.; Zhang, G. Nitrogen-doped graphdiyne as a metal-free catalyst for high-performance oxygen reduction reactions. *Nanoscale,* **2014**, *6*(19), 11336-11343.
       [http://dx.doi.org/10.1039/C4NR03185G] [PMID: 25141067]

[55]   Zhao, Y.; Wan, J.; Yao, H.; Zhang, L.; Lin, K.; Wang, L.; Yang, N.; Liu, D.; Song, L.; Zhu, J.; Gu, L.; Liu, L.; Zhao, H.; Li, Y.; Wang, D. Few-layer graphdiyne doped with sp-hybridized nitrogen atoms at acetylenic sites for oxygen reduction electrocatalysis. *Nat. Chem.,* **2018**, *10*(9), 924-931.
       [http://dx.doi.org/10.1038/s41557-018-0100-1] [PMID: 30082882]

[56]   Gao, Y.; Cai, Z.; Wu, X.; Lv, Z.; Wu, P.; Cai, C. Graphdiyne supported Single-Atom-Sized Fe Catalysts for the Oxygen Reduction Reaction: DFT Predictions and Experimental Validations. *ACS Catal.,* **2018**, *8*, 10364-10374.
       [http://dx.doi.org/10.1021/acscatal.8b02360]

[57]   Liu, R.; Liu, H.; Li, Y.; Yi, Y.; Shang, X.; Zhang, S.; Yu, X.; Zhang, S.; Cao, H.; Zhang, G. Nitrogen-doped graphdiyne as a metal-free catalyst for high-performance oxygen reduction reactions. *Nanoscale,* **2014**, *6*(19), 11336-11343.
       [http://dx.doi.org/10.1039/C4NR03185G] [PMID: 25141067]

[58]   Das, B.K.; Sen, D.; Chattopadhyay, K.K. Implications of boron doping on electrocatalytic activities of graphyne and graphdiyne families: a first principles study. *Phys. Chem. Chem. Phys.,* **2016**, *18*(4), 2949-2958.
       [http://dx.doi.org/10.1039/C5CP05768J] [PMID: 26735306]

[59]   Chen, X.; Qiao, Q.; An, L.; Xia, D. Why Do Boron and Nitrogen Doped a- and g-Graphyne Exhibit Different Oxygen Reduction Mechanism: A First-Principles Study. *J. Phys. Chem. C,* **2015**, *119*, 11493-11498.
       [http://dx.doi.org/10.1021/acs.jpcc.5b02505]

[60]   Zhang, S.; Cai, Y.; He, H.; Zhang, Y.; Liu, R.; Cao, H.; Wang, M.; Liu, J.; Zhang, G.; Li, Y.; Liu, H.; Li, B. Heteroatom doped graphdiyne as efficient metal-free electrocatalyst for oxygen reduction

reaction in alkaline medium. *J. Mater. Chem. A Mater. Energy Sustain.,* **2016**, *4*, 4738-4744.
[http://dx.doi.org/10.1039/C5TA10579J]

[61]   Zhang, J.; Feng, X. Graphdiyne Electrocatalyst. *Joule,* **2018**, *2*, 1396-1409.
[http://dx.doi.org/10.1016/j.joule.2018.07.031]

[62]   Wang, X.; Yang, Z.; Si, W.; Shen, X.; Li, X.; Li, R.; Lv, Q.; Wang, N.; Huang, C. Cobalt-nitroge--doped graphdiyne as an efficient bifunctional catalyst for oxygen reduction and hydrogen evolution reactions. *Carbon,* **2019**, *147*, 9-18.
[http://dx.doi.org/10.1016/j.carbon.2019.02.033]

[63]   Guo, Y.; Liu, J.; Yang, Q.; Ma, L.; Zhao, Y.; Huang, Z.; Li, X.; Dong, B.; Fu, X.Z.; Zhi, C. Metal–tuned acetylene linkages in hydrogen substituted graphdiyne boosting the electrochemical oxygen reduction. *Small,* **2020**, *16*(10), e1907341.
[http://dx.doi.org/10.1002/smll.201907341] [PMID: 32049440]

[64]   Chen, X.; Ong, W-J.; Kong, Z.; Zhao, X.; Li, N. Probing the active sites of site-specific nitrogen doping in metal-free graphdiyne for electrochemical oxygen reduction reactions. *Sci. Bull. (Beijing),* **2020**, *65*, 45-54.
[http://dx.doi.org/10.1016/j.scib.2019.10.016]

CHAPTER 4

Graphene as a Support in Heterogeneous Catalysis

Dipika Konwar[1], Prantika Bhattacharjee[1] and Utpal Bora[1,*]

[1] *Department of Chemical Sciences, Tezpur University, Napaam, Tezpur, Assam, 784028, India*

Abstract: This chapter provides a brief outline of the characteristics and applications of graphene as a support in different heterogeneous catalytic processes. Graphene, an allotrope of carbon having two-dimensional hexagonal arrangements, is an easily available and low-cost support for various catalytic conversions. The high surface area, thermal and chemical stability, lattice arrangement, defects, *etc.,* are the fundamental structural contributions of graphene as support in heterogeneous catalysis. Synthesis of graphene and binding of one or more metal nanoparticles on it has been achieved *via* different well-established methodologies that are highlighted here. Modifications in graphene with numerous physical and chemical methods have enhanced its overall activity and selectivity as support.

Keywords: Bimetallic, Catalyst support, Doping, Electrocatalyst, Heterogenous catalysis, Nanocomposite, Selectivity.

INTRODUCTION

The heterogeneous catalytic process developed by scientists in the first half of the 18th century plays a significant role in the modern-day world, from small-scale syntheses to industrial applications [1]. In a heterogeneous catalytic system, the catalyst occupies a different phase from the reactants, which facilitates the smooth progression of the catalytic reaction without itself being consumed throughout the process. The key component in a heterogeneous catalysis is the active site of the catalyst [2]. A comprehensive understanding of the structure and composition of the active site is crucial for identifying the properties of heterogeneous catalysts [3]. About 90% of the total industrial processes rely on heterogeneous catalysis due to high selectivity, reusability of the catalyst, and easy separation of products [4]. In this type of catalysis, the catalyst is deposited on an inert solid support (such as silica, alumina, zeolite, carbon materials, *etc.*), having a high surface area [5, 6].

---

[*] **Corresponding author Utpal Bora:** Department of Chemical Sciences, Tezpur University, Napaam, Tezpur, Assam, 784028, India; E-mail: ubora@tezu.ernet.in & utbora@yahoo.co.in

Manorama Singh, Vijai K. Rai and Ankita Rai (Eds)

The porous nature and high surface area of the solid supports help in higher dispersion of the catalysts on the surface, thereby increasing the activity of the catalyst [6 - 8]. The use of solid supports in nanoparticle-catalyzed reactions also prevents the agglomeration of nanoparticles, thereby maintaining the catalytic activity [9, 10]. Out of the different solid supports used, carbon-based materials are the most common in heterogeneous catalysis [5, 11]. The chemically inert nature of carbon-based materials results in a high surface area, porosity, and thermal stability, which promote the higher activity and selectivity of the catalyst [11]. Carbon supports have a wide range of applications in the field of organic and inorganic synthesis, electrocatalysis of fuel cells, solar cells, sensors, *etc* [12 - 17]. Graphene, an allotrope of carbon, has a two-dimensional layered structure in which the sp^2 hybridized carbon atoms (Fig. **1**) are arranged in a hexagonal pattern, which has been successfully used as a support in heterogeneous catalysis [4, 5, 18]. Due to the sp^2 hybridization, one of the unhybridized p-orbitals overlaps with the neighboring carbon atom and forms π-bonds. These π-orbitals are present above and below the hexagonal sheet of graphene and thus provide excellent electrical conductivity [19]. Moreover, graphene has several distinctive properties such as high electron mobility, quantum hall effect, thermal conductivity at room temperature, well-arranged atomic lattice, chemical and thermal stability, *etc* [4, 18, 20]. The arrangement of atomic lattice in graphene is such that the charged particles can move through the inter-planar spaces without any scattering. These properties indicate that graphene can act as a better channel for faster electrochemical operations with low power utilization [19].

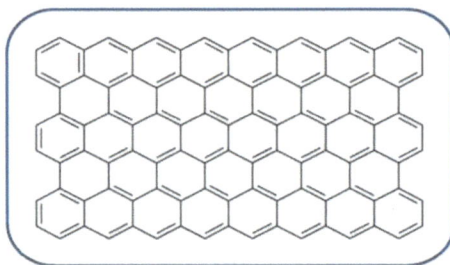

Fig. (1). 2-D hexagonal sheet structure of graphene.

Two-dimensional sheet structure, along with single-atom thickness, results in a high surface-to-volume ratio of graphene [5, 21]. This large surface area of graphene leads to the binding of metals, metal-oxides, and nanoparticles on its surface and facilitates the charge transfer either from graphene to the metal or vice-versa in different redox processes. The high surface area and two-dimensional structure of graphene allow the adsorption of various molecules on its surface. The binding of these adsorbates changes the electrical conductivity of graphene, showing its sensing ability to different adsorbate molecules [22].

Functionalization of the graphene layer with electronically diverse functional groups enhances the chemical stability, solubility, and dispersibility of the support [5, 23]. This functionalization can be done through covalent and non-covalent interactions. In covalent functionalization, the generation of electrical charges induces the electrostatic interactions within the graphene sheets. On the other hand, non-covalent interaction is the physical adsorption of molecules on the graphene that breaks the symmetry of the layered structure [24]. The presence of structural defects in graphene is responsible for the strong binding interactions with the metal catalysts [20]. These attractive properties of graphene make it promising support in different areas of heterogeneous catalysis. Not only graphene but other graphene-based materials such as graphene oxide, reduced graphene oxide, doped graphene are also used in heterogeneous catalysis. The binding of oxygenated species on graphene enhances the electron transfer, thereby boosting its electrochemical behaviour [24].

SYNTHESIS AND DESIGNING OF GRAPHENE AS SUPPORT

From the literature, it is seen that the first production of graphene was achieved by the micromechanical cleavage method from graphite *via* extraction [25, 26]. But the major drawbacks in the micromechanical cleavage technique demanded the development of more feasible methods for the large-scale synthesis of graphene. Following, in 2004, Novoselov *et al.* successfully synthesized graphene by mechanical exfoliation from graphite [27]. Some other techniques, including chemical exfoliation, thermal chemical vapour deposition, chemical synthesis, microwave synthesis, plasma-enhanced chemical vapour deposition, unzipping carbon nanotubes, *etc.*, have been used in recent years for the synthesis of graphene [5, 26, 27]. Designing of graphene by covalent (such as direct doping, residual functionalities) and non-covalent modifications (such as electrostatic interaction, hydrogen bonding, π-π stacking interaction) have improved its stability, solubility, electronic, photoelectronic, mechanical, and thermal properties [28, 29]. Moreover, the surface area of graphene support can be increased by thermolytic cracking, self-assembly, *etc.*, which leads to the binding of a large number of catalysts on its surface [5].

METAL AND METAL OXIDE CATALYSTS ON GRAPHENE SUPPORT

Palladium on graphene: Palladium (Pd), a highly abundant noble metal, has prominent catalytic activity among the different metals. Pd on graphene catalyst is widely used in carbon-carbon and carbon-heteroatom bond formation reactions. Li *et al.* in 2010 reported the Suzuki-coupling reaction (Scheme **1**) using Pd nanoparticle-graphene hybrid as catalyst [30]. They synthesized graphene by heating graphene oxide with sodium hydroxide solution. In this methodology, Pd nanoparticles are homogenously dispersed on the modified graphene support with a controllable size of nanoparticles, generating the Pd-graphene hybrid catalyst with high efficiency to carry out the Suzuki reaction in an aqueous medium. The surfactant sodium dodecyl sulphate (SDS) acts as a reducing agent for the synthesis of Pd nanoparticles from $Pd(OAc)_2$. This Pd-graphene hybrid has very high dispersibility and can be re-used up to the 10th cycle easily.

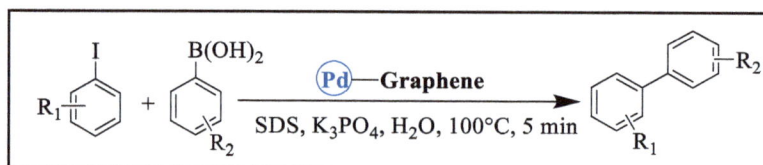

Scheme (1). Synthesis of biaryls using supported Pd nano-catalyst.

Pd on graphene support can be used for the oxidation and reduction reactions of carbonyl compounds and alcohols. In 2012, Li *et al.* reported the Pd on graphene support catalyzed oxidation of carbon monoxide (CO) [31]. In this oxidation process, graphene was synthesized by the Hummers method and thermal reduction process from graphite powder. Tetrachloropalladinic acid (H_2PdCl_4) was used as the Pd source for the preparation of graphene-supported Pd catalyst. In this oxidation process, initially, CO and molecular oxygen were chemisorbed on the catalyst surface and then reacted together. This oxidation followed the Langmuir-Hinshelwood mechanism. The turnover frequency (TOF) of this catalytic CO oxidation increases with increasing the amount of Pd loading.

Graphene-supported Pd can also act as a sensor, and this was reported by Wolfbeis and co-workers where they used Pd-graphene nanocomposites for hydrogen gas (H_2) sensing purposes [32]. A layer-by-layer deposition method was applied for the synthesis of this nanocomposite. When the nanocomposite was exposed to H_2 gas, a sharp decrease in the resistance was observed, which might be due to dissociation as well as dissolution of H_2 on metal nanoparticles. This diminishing nature of resistance confirmed the transfer of electrons to graphene in the presence of H_2 gas. When a Pd-free condition was applied, graphene was shown to be very less sensitive to hydrogen. However, the layer-by-layer deposition of metals on it resulted in a chemo-sensitive material. This nanocomposite can also be used for the detection of NO_2 and humidity in the air.

Gold on graphene: Graphene-supported gold (Au) catalysts are fruitful tools for the various oxidation-reduction processes. In 2016, Valiollahi *et al.* reported the electro-oxidation of sodium borohydride ($NaBH_4$) catalyzed by gold on graphene support [33]. In their methodology, they used gold nano-cage immobilized on the graphene nano-sheet for electro-oxidation and studied the phenomenon through cyclic voltammetry. The porosity and hollow nature of the synthesized catalyst are accountable for the stability and electrocatalytic activity.

Copper on graphene: Copper (Cu) is an easily available and low-cost metal catalyst itself or in supported form for a variety of chemical reactions. Copper nano-particles on graphene supports are an effective photocatalyst for coupling of nitro aromatics to azo compounds irradiated in visible light. This photocatalytic conversion was reported by Guo *et al.* in 2014 (Scheme **2**), where Cu nano-particles were in a metallic state instead of the oxide form [34]. Through the surface plasmon resonance effect, nanoparticles absorbed energy from the visible light that excited the bound electrons. These excited electrons at the surface of the catalyst facilitated the breaking of N-O bonds of nitro compounds and furnished the corresponding azo compounds.

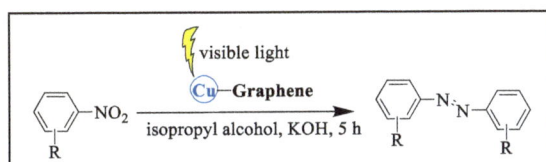

Scheme (2). Visible-light mediated nitro to azo conversion.

Furthermore, the oxidation of CO can be done with a Cu-graphene hybrid catalyst with high efficiency and at a low cost [35]. The high catalytic activity is due to resonance among the electronic states of participating atoms.

Cobalt/Cobalt oxide on graphene: The production of hydrogen as a sustainable and clean energy source through water splitting by an electrocatalyst is the major demand in today's world. The water splitting process involves the oxygen and hydrogen evolution reactions at the electrodes. In 2018, Li *et al.* reported the graphene-supported cobalt (Co) and nano-cobalt oxide (Co_3O_4) electrocatalyst for oxygen evolution reaction (OER) with high catalytic efficiency [36]. By using electrochemical exfoliation process, graphene was synthesized, and further electrodeposition of the catalyst was done. The kinetics of OER was determined with the help of the Tafel polarization plot. The evolution of oxygen gas was observed on the surface of the rotating disk electrode (RDE) in bubble form. The synergistic effect of atomic Co and nano-Co-oxide improves the catalytic activity of this electrochemical reaction.

BIMETALLIC CATALYSTS ON GRAPHENE SUPPORT

Platinum-Ruthenium on graphene: Bimetallic Pt-Ru nanoparticles supported on graphene is an effective catalyst for the oxidation reaction of methanol. Bo and co-workers reported the use of this bimetallic catalyst for oxidation of methanol, where graphene was synthesized through microwave-assisted plasma-enhanced chemical vapour deposition process [37]. The binding of Pt-Ru nanocatalysts on graphene support was done by co-electrodeposition under a nitrogen atmosphere. They observed the methanol oxidation reaction through cyclic voltammetry and Tafel analysis. This catalytic system is well-controllable in terms of deposition of nanoparticles along with the dispersibility of the support.

Palladium-Gold on graphene: Graphene supported Pd-Au bimetallic catalyst finds broad applications in the oxidation of alcohols. The addition of Au into Pd enhances the activity of this hybrid catalyst system. Synthesis of methyl formate *via* oxidation of methanol in the presence of Pd-Au nanoparticles on graphene support was reported by Wang *et al.* in 2013 [38]. This catalytic conversion is highly efficient with 100% selectivity. Here synthesis of graphene was done by vacuum promoted thermal expansion method from graphene oxide. Uniform dispersion of the bimetallic catalyst on the support was carried out using the deposition-precipitation method. In this bimetallic system, the nanoparticles had a twin-particle structure, and Au acted as a promoter for the Pd anchored to the support. The electrical conductivity of graphene promoted the charge exchange between the two metals and also donated its π-electrons to the nanoparticles that reinforced the synergistic effect of Au and Pd and improved the catalytic performance. The significant properties of this Pd-Au catalyst on graphene make it effective for the reduction of nitrophenol. This was done by Liu *et al.,* where thermal exfoliation along with hydrogen reduction methods was used for the synthesis of graphene [39]. The reduction process was monitored by UV-vis spectroscopy. Self-assembly of Pd-Au nanoparticles on the support was expected to stabilize the catalyst system.

Platinum-Cobalt on graphene: Composite of Pt-Co nanocatalyst on graphene support has higher activity as compared to that of atomic Pt nanoparticles. In 2014, Kargupta and co-workers reported a cost-effective hydrogen production reaction from hydrolysis of $NaBH_4$ catalyzed by graphene-supported Pt-Co nanocomposite [40]. With the help of a control valve in the reactor, the aqueous solution of $NaBH_4$ and catalyst were mixed, and the evolved H_2 gas was collected after the removal of water by using a silica drier. This nanocomposite actually controls the generation of hydrogen gas from the $NaBH_4$ solution. Effective interaction and well-dispersibility with graphene support stabilize these Pt-Co nanoparticles, enhancing hydrogen production.

Cobalt-Nickel on graphene: Bimetallic Co-Ni nanocatalyst is a low-cost non-noble hybrid catalyst for catalytic dehydrogenation reactions. Feng *et al.* reported the dehydrogenation reaction of amine borane using this graphene-supported magnetic nanocatalyst [41]. In this catalytic system, graphene appears to be the most prominent support as compared to silica, alumina, and carbon black. Better stability, dispersibility, and magnetic reusability are the major advantages of this graphene-supported bimetallic catalyst for hydrolytic dehydrogenation of ammonia borane.

MODIFIED GRAPHENE SUPPORT

Reduced graphene oxide support: Reduced graphene oxide is a significant nanomaterial that is extensively used in diverse catalytic conversions. Although it is an electrically insulating material, its conductive nature can be restored through the reduction process [5]. Different reducing agents, such as sodium borohydride, hydrazine, hydroquinone, ethylenediamine, *etc.,* are used for the preparation of reduced graphene oxide [42]. Instead of applying chemical agents, reduced graphene oxide can also be synthesized by other methods like photocatalytic, thermal, microwave, sonochemical treatment, *etc* [42]. Depending upon the nature of reduction routes, variation in stability, dispersibility, structural defects, and conductivity of synthesized graphene is observed.

Mu and co-workers reported the reduced graphene oxide supported Pt catalyst for oxygen reduction reaction where the presence of oxygenated species on the support stabilised the metal catalyst and enhanced the catalytic activity [43]. They synthesized the reduced graphene oxide through ethanol and hydrogen reduction methods and the hydrophilic nature of the same increased the homogeneous dispersibility of the metal catalyst. The oxygen-containing groups on the surface also increased the interactions of the Pt and the support. This oxygen reduction reaction was monitored through the cyclic voltammetry method.

Bimetallic Pd-Ag nanoparticles supported on the reduced graphene oxide is an effective nanocomposite for the production of hydrogen gas which was carried out by Xu and co-workers [44]. In this hydrogen generation method, the bimetallic catalysts were immobilized on the zirconia-porous carbon metal-organic framework that was supported on reduced graphene oxide. The introduction of zirconia allowed the transfer of electrons to Pd-Ag and thereby changed the electronic structure of these bimetallic nanoparticles. The electron-rich surface of the nanoparticles is responsible for the breakdown of formic acid to form nanoparticle-formate intermediate, which is again cleaved to form Pd-hydride species. This Pd-hydride species reacts with a proton to generate H_2 gas. The synergistic effect between Pd-Ag nanoparticles and the support promoted the electrocatalytic activity of the nanocomposite. Another type of reduced graphene oxide-supported hydrogen generation method was reported by Ali and co-workers under photocatalytic conditions promoted by visible light [45]. Here sulphides of cadmium (CdS) and molybdenum (MoS$_2$) were supported on the reduced graphene oxide *via* hydrothermal process. The electronic charge generated in the surface of CdS due to photo-irradiation was shifted to MoS$_2$ through reduced graphene oxide, and thus H$_2$ gas is evolved from water. The intimate interactions of the two metal sulphides on the support improved the photocatalytic activity for hydrogen evolution reaction in an aqueous medium. The phosphide of Mo supported on reduced graphene is also an efficient catalyst for the hydrogen evolution reaction [46]. This non-noble metal phosphide is capable of existing in both acidic and basic media with prominent catalytic activity. Moreover, reduced graphene supported Pt [47], nickel ferrite [48] nanomaterials are other examples of electrocatalysts for hydrogen evolution reactions.

The catalytic reduction of nitrogen oxide can be done by using reduced graphene oxide support. Ye *et al.* carried out this type of reduction reaction in the presence of Mn-Ce on TiO$_2$, supported over reduced graphene oxide [49]. The thermal reduction method was used for the preparation of reduced graphene oxide, where metal nanoparticles were dispersed by the impregnation process. The reduction reaction was performed in a fixed bed reactor in which a supported metal catalyst was placed. Mass flow controller was used for the controlled release of nitrogen oxide, ammonia, and oxygen gas to the reactor. The adsorption of ammonia on the active sites of the support could selectively reduce nitrogen oxide to N$_2$ gas and water. The amount of nitrogen oxide introduced and released in the reaction was measured with the Chemi-luminescent detection analyzer, and hence the efficiency of the reaction could be determined. Due to the presence of oxygen-containing groups in the support, the metal nanoparticles bind very strongly and display higher catalytic activity. The large surface area of the support results in the good dispersibility of the metal catalyst.

Moreover, reduced graphene oxide can be utilized as support for the cross-coupling reactions in the presence of metal catalysts. Mahanta *et al.* reported the Sonogashira coupling reaction (Scheme **3**) catalyzed by Pd nanoparticles supported on reduced graphene oxide without using any ligand at room temperature under air [50]. Hydrogen reduction technique was utilized for the preparation of reduced graphene oxide supported metal composite. The evenly deposited metal nanoparticles on the support carried on the reaction efficiently with excellent yield (up to 98%) and could be re-used up to sixth cycle easily.

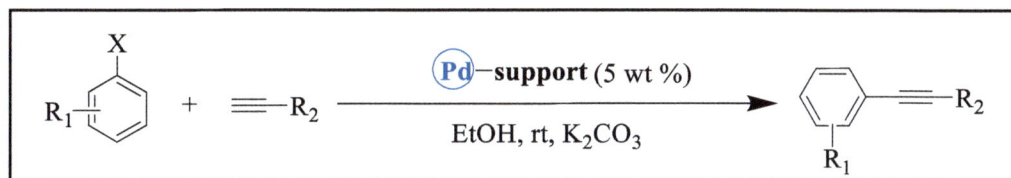

Scheme (3). Sonogashira coupling of aryl halide and phenylacetylene derivatives.

Porous graphene support: The name porous graphene itself indicates the presence of some pores or holes within the structure. It is actually a modified form of graphene where holes are present inside the atomic planes due to the absence of carbon atoms [51, 52]. The presence of pores in graphene induces bandgap opening, which modifies the electronic properties in porous graphene. Generally, porous materials are very useful in catalysis, separation process, sensing, energy storage, *etc* [51, 52]. The synthesis of porous graphene is done by chemical and physical etching, acid etching, carbothermic reduction, *etc* [51 - 53]. In the chemical etching process, some oxidising agents such as H_2O_2, O_2, MnO_2, Fe_2O_3, *etc.*, are used for the creation of vacancies or holes. Photoelectric beam, O_2 plasma are used in the physical etching method. In the acid etching process, pores are generated by using nitric and phosphoric acids. Along with these processes, some other methods, including partial combustion, electrochemical exfoliation, microwave method, laser synthesis, *etc.*, are used for the synthesis of porous graphene [52]. The porous nature of this support increases the surface area that leads to well-dispersed metal catalysts on it. Moreover, the porosity improves the mechanical, chemical, and thermal stability of the support. The sponge-like structure of porous graphene helps in ion transport, sieving, energy storage process, *etc.*

Qiu *et al.* reported the Pt nanoparticles on porous graphene support as an electrocatalyst for the oxidation of methanol with very high catalytic efficiency [54]. The metal nanoparticles were synthesized by using a ferritin template, and the binding of the metal catalyst on the support was done with the electrodeposition method. Ferritin template acted as a size controlling agent for the nanoparticles in this methanol oxidation reaction. The catalytic process was observed through the cyclic voltammetry technique. The interconnected pores of the support resulted in the easy flow of methanol during the electrocatalytic process.

Porous graphene can be used as support in metal-catalyzed oxygen evolution reactions. Xia *et al.* reported this type of reaction where porous graphene was used as support for $Au\text{-}NiCo_2O_4$ catalyst [55]. The support was prepared by the ion exchange-activation combination process in one step. The interactions of the support with $NiCo_2O_4$ improved the electrical conductivity of $NiCo_2O_4$. Highly electronegative metal Au adsorbed electrons and stabilized the higher oxidation states of nickel and cobalt ions which are the active centres for oxygen evolution. This conductive support provides better stability, activity, and large surface area for the oxygen evolution reaction.

Doped graphene support: Doping improves the electrocatalytic properties of graphene as support. Doping is one of the most effective methods to regulate the semiconducting behaviour of graphene [56]. The doping of graphene can be done by using hetero atom, chemical modification, and tuning of the electrostatic field [56, 57]. The term hetero atom doping itself indicates the insertion of heteroatoms in place of carbon atoms of graphene. When a hetero atom is inserted into the sp^2 hybridized hexagonal network, electroneutrality between dopant and carbon atoms breaks down and generates charged sites that are favourable for adsorption of the metal catalyst. The most commonly used hetero atoms for this doping process are boron (B) and nitrogen (N) because of their similarity in atomic size with that of the carbon atom. The aim of insertion of the heteroatom is to open the bandgap, and the presence of Fermi levels in conduction and valence band results in the enhancement of semiconducting properties. This type of doping leads to the use of graphene in electronic devices. Chemical vapour deposition, ion-irradiation, arc discharge, electrothermal reaction, segregation growth, and solvothermal methods can be used for the heteroatom doping of graphene. Similar to heteroatom doping, chemical modifications are also done to open the bandgap of graphene. In the chemical modification, NO_2 and NH_3 are used for donor-acceptor interactions of electrons and holes. Here tuning of Fermi level is also done. Organic molecules get a better preference for this type of doping. The electrostatic field tuning doping method involves the control of the magnetic and electronic properties of graphene. Here the bandgap of graphene cannot be opened.

In the case of hetero atom doped graphene, the nitrogen atom is preferably used for doping. Doping of nitrogen leads to three main bonding types, namely graphitic or quaternary, pyrrolic, and pyridinic N configurations (Fig. **2**) [58]. Graphitic N configuration indicates that the sp^2 hybridized carbon atoms are replaced by nitrogen atoms in the six-membered ring of graphene. Similar to the graphitic N, pyridinic N is also sp^2 hybridized, and this N atom is attached to a single oxygen atom along with the other two carbon atoms of the hexagonal ring. In five-membered pyrrolic N bonding, the nitrogen atom is sp^3 hybridized and donates its p-electrons to the ring system [56, 57]. The distribution of charge and density of spin in carbon atoms of graphene are influenced by the presence of dopant atoms [59, 60]. As a result, some active regions are generated in the graphene support that helps in the catalytic processes. This N-doped graphene is widely used as support for different electrocatalytic conversion reactions.

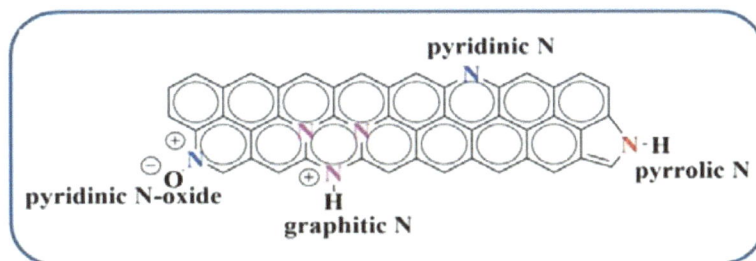

Fig. (2). Three different types of N-doped graphene bonding configurations.

Xiong *et al.* reported the Pt nanoparticles catalyzed N-doped graphene support for oxidation of methanol [61]. They prepared the N-doped graphene from graphene oxide *via* a thermal annealing process in ammonia. This conductive N-doped graphene was used as support for the Pt nanoparticles. The deposition of nanoparticles was done by the aqueous electroless deposition method. The controllable size of nanoparticles, uniform dispersion on the support, and N-doping were all favourable factors for the electrocatalytic oxidation reaction of methanol. The catalytic process was monitored through cyclic voltammetry.

Yu and co-workers carried out the water oxidation reaction to generate the oxygen gas using a nano $CoSe_2$ catalyst supported on N-doped graphene [62]. The hydrothermal reduction method was applied for the binding of nano metal particles on the support. These cobalt selenide metal nanoparticles are less-expensive and show better stability. The high oxidation state of cobalt ion elevated the electrophilicity of adsorbed oxygen, facilitating the generation of hydroperoxide ion, which finally furnished molecular oxygen. The generation of oxygen in this reaction was analysed from the Tafel plot, and they observed that this nanocomposite had the smallest value of the Tafel slope that proved the high efficiency of the electrocatalyst. Similar to that of cobalt selenides, nickel nanoparticles supported on N-doped graphene can also be used for the water-splitting process for the generation of oxygen. This was reported by Xu and co-workers in 2017 [63]. In their methodology, they had used a Ni-based metal-organic framework (MOF) that was anchored to the support through the annealing process. The transfer of electrons from Ni core to graphene altered the electronic properties of N-doped graphene and improved the electrocatalytic activity. The easy and cost-effective synthesis of this Ni-MOF catalyst supported on N-doped graphene with longevity and high catalytic activity makes the nanocomposite a promising electrocatalyst for the water-splitting process. Not only metallic Ni but also the sulfide of Ni is fruitful for oxygen and hydrogen evolution processes. Jayaramulu *et al.* synthesized a Ni_7S_6 supported catalyst on N-doped graphene *via* a solvothermal process that showed very high stability and immense efficiency in the alkaline medium for both hydrogen and oxygen evolution reactions [64].

Ramaprabhu and co-workers carried out the oxygen reduction reaction using Pt nanoparticles supported on N-doped graphene [65]. The support was prepared by nitrogen plasma treatment, and further chemical reduction was done for the deposition of metal nanoparticles. The plasma treatment generated the pyrrolic N defects that resulted in better binding sites for the nanoparticles. The close interactions of the catalyst with support and its high conductivity due to doping enhanced the catalytic oxygen reduction process in fuel cells. Another type of N-doped graphene-supported catalyst for oxygen reduction reaction was reported by Wu *et al.*, where they had used iron oxide nanoparticles as catalysts [66]. Hydrothermal and thermal techniques were applied for the synthesis of Fe_3O_4 on N-doped composite. The well-dispersibility of the nanoparticles on the large surface of the support and durability of the electrocatalyst on alkaline solution increases the oxygen reduction reaction.

There are some examples available in the literature for the B-doped graphene as supports in different catalytic reactions. In B-doped graphene, the re-distribution of electronic charges occurred because of the low electron density of boron as compared to that of neighbouring carbon atoms. This re-distribution of charge particles enhances the conductivity of doped graphene and makes it an excellent material for electrocatalytic reactions. Sun *et al.* reported the electrooxidation reaction of methanol to form carbon dioxide and water using Pt nanoparticles on the B-doped graphene support [67]. They prepared the support by thermal annealing technique, and nanoparticles were deposited in the defect sites generated due to B-doping. The stability and catalytic process were observed through cyclic voltammetry. The low electronegative B atom donated its electrons to the Pt atom, and this interaction between nanoparticles and the support resulted in effective oxidation of methanol. Decomposition of methanol to form carbon monoxide and hydrogen gas using bimetallic Ru-Pt catalyst on B-doped graphene support was reported by Jiang and co-workers through theoretical analysis [68]. They used bimetallic catalyst in order to improve the surface of Pt, as the formation of a small amount of carbon monoxide could poison it. With the help of theoretical calculations, they found that the decomposition of methanol to carbon monoxide generates methoxy and aldehyde intermediates.

Iridium oxide nanoparticles on B-doped graphene support as electrocatalyst for oxygen evolution reaction was reported by Joshi *et al.* in 2020 [69]. The doping of B atom on graphene support was done by pyrolysis, and nanoparticles were deposited through hydrothermal technique. The synergistic effect of nanoparticles and B-doped graphene support stabilized the electrocatalyst and increased the durability in acidic medium, and hence improving the oxygen evolution from water splitting. The catalytic activity of the reaction was determined using a rotating disk electrode through linear sweep voltammetry.

Wang and co-workers carried out the selective reduction of nitric oxide with H_2 gas in the presence of Pt nanoparticles supported on B-doped graphene [70]. Hydrothermal synthesis was done for the preparation of a B-doped graphene-supported Pt nanocomposite. They carried out the reaction in a continuous fixed bed reactor and calculated the data for the concentration of nitric oxide used and released in the process, which proved the occurrence of reduction. The controllable size of nanoparticles, uniform dispersion on the support, presence of defects due to boron doping, chemisorptions of H_2 gas and nitric oxide on the support are all favourable factors for this selective catalytic reduction process.

ROLE OF GRAPHENE AS SUPPORT IN HETEROGENEOUS COUPLING REACTIONS

While the primary role of supports in catalysis is to anchor metal particles, they can have more than one function to play [4]. The high catalytic activity of graphene is connected to its outstanding electrical conductivity. To get an insight into how the graphene support works in heterogeneous catalysis, it is important to understand the catalyst-support interactions. In any heterogeneous catalytic coupling reactions, the elementary steps contain oxidative addition, transmetallation, and reductive elimination involving charge donation and acceptance throughout the catalytic pathway [71].

Yang *et al.* has tried to rationalize the fact by studying the electronic interactions between the metal particles and conducting graphene support for a Pd-graphene oxide catalyst acting in Suzuki coupling reaction of 4-bromobenzoic acid and phenylboronic acid (Scheme **4**) [72]. In the oxidative addition step, the Br-C bond of 4-bromobenzoic acid is cleaved by the Pd-graphene catalyst. This step requires charge donation from the catalyst-support complex to the reactant. During the transmetallation step, charge flows from the Pd-graphene cluster towards the phenyl boronic acid substrate when the C-B bond breaks and again flows back when forming the Pd-C bond. Thus, this step requires both charge donation and back donation. Next, in the reductive elimination step, the C-C bond is formed, and the product is released from the catalyst surface. In this process, the catalyst readily accepts the charge. Hence, it is clear that the presence of conductive support like graphene assists both the charge donation and withdrawal process simultaneously through the catalytic cycle. This dual-activation nature of graphene-supported catalysts makes it applicable to a diverse range of heterogeneous catalyzed reactions.

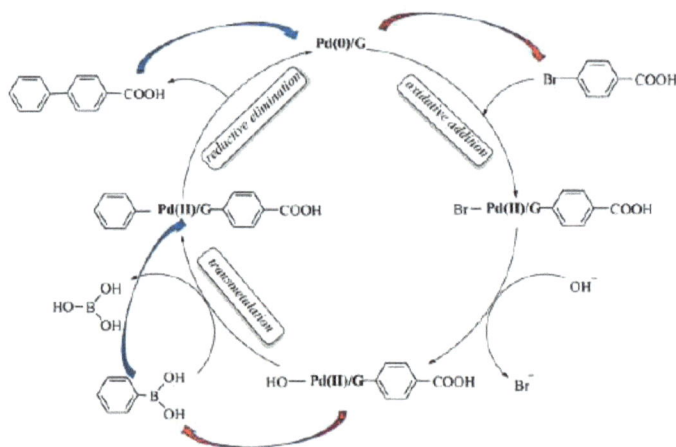

Scheme (4). Mechanism of Suzuki coupling reaction catalyzed by Pd-graphene catalyst showing the charge flow between the Pd-graphene complex and the coupling partners.

CONCLUSIONS

Graphene as support in heterogeneous catalysis exhibits excellent catalytic activity and selectivity. In this chapter, we endeavoured to highlight the applications of graphene as support in diverse catalytic processes. The attractive physical and chemical properties of graphene have encouraged researchers to make tremendous progress in the field of heterogeneous catalysis, considering graphene as support. The high surface area and tunable electrical properties of the support help the metal nanoparticles anchor on the graphene surface strongly. This stronger interaction prevents metal leaching, shaping the metal-graphene complex as a true heterogeneous catalytic species. Graphene-supported metal nanoparticles find wide applications in the field of cross-coupling chemistry that lead to the formation of carbon-carbon and carbon-heteroatom bonds. These functionalization processes have potential applications in small-scale synthesis as well as in material and pharmaceutical sciences. We also tried to discuss the graphene-supported photocatalysts and electrocatalysts that are commonly used in oxygen and hydrogen evolution reactions. Depending upon the presence of either single metal or bi-metal nanoparticles on the support, there lie variations in the catalytic properties as well. Modifications of the graphene support help in tuning the physical and chemical properties, some of which are also highlighted here. Thus, as less-expensive and highly active support, graphene has the potential to provide more fruitful applications in academic and industrial researches in the near future.

CONSENT FOR PUBLICATION

Not applicable.

CONFLICT OF INTEREST

The author declares no conflict of interest, financial or otherwise.

ACKNOWLEDGEMENTS

The authors gratefully acknowledge Tezpur University, Assam and SERB, New Delhi, for the financial support (No. EMR/2016/005944) and DST-Govt. of India for providing INSPIRE fellowship to DK.

REFERENCES

[1]   George, S.M. Introduction: heterogeneous catalysis. *Chem. Rev.,* **1995**, *95*, 475-476.
      [http://dx.doi.org/10.1021/cr00035a001]

[2]   Christensen, C.H.; Nørskov, J.K. A molecular view of heterogeneous catalysis. *J. Chem. Phys.,* **2008**, *128*(18), 182503.
      [http://dx.doi.org/10.1063/1.2839299] [PMID: 18532788]

[3]   Schlögl, R. Heterogeneous catalysis. *Angew. Chem. Int. Ed. Engl.,* **2015**, *54*(11), 3465-3520.
      [http://dx.doi.org/10.1002/anie.201410738] [PMID: 25693734]

[4]   Fan, X.; Zhang, G.; Zhang, F. Multiple roles of graphene in heterogeneous catalysis. *Chem. Soc. Rev.,* **2015**, *44*(10), 3023-3035.
      [http://dx.doi.org/10.1039/C5CS00094G] [PMID: 25777748]

[5]   Julkapli, N.M.; Bagheri, S. Graphene supported heterogeneous catalysts: an overview. *Int. J. Hydrogen Energy,* **2015**, *40*, 948-979.
      [http://dx.doi.org/10.1016/j.ijhydene.2014.10.129]

[6]   Xia, Q.H.; Ge, H.Q.; Ye, C.P.; Liu, Z.M.; Su, K.X. Advances in homogeneous and heterogeneous catalytic asymmetric epoxidation. *Chem. Rev.,* **2005**, *105*(5), 1603-1662.
      [http://dx.doi.org/10.1021/cr0406458] [PMID: 15884785]

[7]   Armbrüster, M.; Kovnir, K.; Friedrich, M.; Teschner, D.; Wowsnick, G.; Hahne, M.; Gille, P.; Szentmiklósi, L.; Feuerbacher, M.; Heggen, M.; Girgsdies, F.; Rosenthal, D.; Schlögl, R.; Grin, Y. Al13Fe4 as a low-cost alternative for palladium in heterogeneous hydrogenation. *Nat. Mater.,* **2012**, *11*(8), 690-693.
      [http://dx.doi.org/10.1038/nmat3347] [PMID: 22683821]

[8]   Lee, I.; Albiter, M.A.; Zhang, Q.; Ge, J.; Yin, Y.; Zaera, F. New nanostructured heterogeneous catalysts with increased selectivity and stability. *Phys. Chem. Chem. Phys.,* **2011**, *13*(7), 2449-2456.
      [http://dx.doi.org/10.1039/C0CP01688H] [PMID: 21103527]

[9]   Holm, M.S.; Saravanamurugan, S.; Taarning, E. Conversion of sugars to lactic acid derivatives using heterogeneous zeotype catalysts. *Science,* **2010**, *328*(5978), 602-605.
      [http://dx.doi.org/10.1126/science.1183990] [PMID: 20431010]

[10] Kesavan, L.; Tiruvalam, R.; Ab Rahim, M.H.; bin Saiman, M.I.; Enache, D.I.; Jenkins, R.L.; Dimitratos, N.; Lopez-Sanchez, J.A.; Taylor, S.H.; Knight, D.W.; Kiely, C.J.; Hutchings, G.J. Solvent-

free oxidation of primary carbon-hydrogen bonds in toluene using Au-Pd alloy nanoparticles. *Science,* **2011**, *331*(6014), 195-199.
[http://dx.doi.org/10.1126/science.1198458] [PMID: 21233383]

[11]   Rodriguez-Reinoso, F. The role of carbon materials in heterogeneous catalysis. *Carbon,* **1998**, *36*, 159-175.
[http://dx.doi.org/10.1016/S0008-6223(97)00173-5]

[12]   Botos, A.; Biskupek, J.; Chamberlain, T.W.; Rance, G.A.; Stoppiello, C.T.; Sloan, J.; Liu, Z.; Suenaga, K.; Kaiser, U.; Khlobystov, A.N. Carbon nanotubes as electrically active nanoreactors for multi-step inorganic synthesis: sequential transformations of molecules to nanoclusters and nanoclusters to nanoribbons. *J. Am. Chem. Soc.,* **2016**, *138*(26), 8175-8183.
[http://dx.doi.org/10.1021/jacs.6b03633] [PMID: 27258384]

[13]   Wang, Z.B.; Zuo, P.J.; Wang, G.J.; Du, C.Y.; Yin, G.P. Effect of Ni on Pt-Ru/C catalyst performance for ethanol electrooxidation in acidic medium. *J. Phys. Chem. C,* **2008**, *112*, 6582-6587.
[http://dx.doi.org/10.1021/jp800249q]

[14]   Tuaev, X.; Rudi, S.; Strasser, P. The impact of the morphology of the carbon support on the activity and stability of nanoparticle fuel cell catalysts. *Catal. Sci. Technol.,* **2016**, *6*, 8276-8288.
[http://dx.doi.org/10.1039/C6CY01679K]

[15]   Kumar, A.; Chaudhary, D.K.; Parvin, S.; Bhattacharyya, S. High performance duckweed-derived carbon support to anchor NiFe electrocatalysts for efficient solar energy driven water splitting. *J. Mater. Chem. A Mater. Energy Sustain.,* **2018**, *6*, 18948-18959.
[http://dx.doi.org/10.1039/C8TA06946H]

[16]   Zhang, Y.; Yun, S.; Wang, C.; Wang, Z.; Han, F.; Si, Y. Bio-based carbon-enhanced tungsten-based bimetal oxides as counter electrodes for dye-sensitized solar cells. *J. Power Sources,* **2019**, *423*, 339-348.
[http://dx.doi.org/10.1016/j.jpowsour.2019.03.054]

[17]   Zheng, W.; Li, Y.; Hu, L.; Lee, L.Y.S. Use of carbon supports with copper ion as a highly sensitive non-enzymatic glucose sensor. *Sens. Actuators B Chem.,* **2019**, *282*, 187-196.
[http://dx.doi.org/10.1016/j.snb.2018.10.164]

[18]   Haag, D.; Kung, H.H. Metal free graphene based catalysts: a review. *Top. Catal.,* **2014**, *57*, 762-773.
[http://dx.doi.org/10.1007/s11244-013-0233-9]

[19]   Brownson, D.A.; Banks, C.E. Graphene electrochemistry: an overview of potential applications. *Analyst (Lond.),* **2010**, *135*(11), 2768-2778.
[http://dx.doi.org/10.1039/c0an00590h] [PMID: 20890532]

[20]   Yam, K.M.; Guo, N.; Jiang, Z.; Li, S.; Zhang, C. Graphene-based heterogeneous catalysis: role of graphene. *Catalysts,* **2020**, *10*, 53.
[http://dx.doi.org/10.3390/catal10010053]

[21]   Navalon, S.; Dhakshinamoorthy, A.; Alvaro, M.; Antonietti, M.; García, H. Active sites on graphene-based materials as metal-free catalysts. *Chem. Soc. Rev.,* **2017**, *46*(15), 4501-4529.
[http://dx.doi.org/10.1039/C7CS00156H] [PMID: 28569912]

[22]   Ambrosi, A.; Chua, C.K.; Bonanni, A.; Pumera, M. Electrochemistry of graphene and related materials. *Chem. Rev.,* **2014**, *114*(14), 7150-7188.
[http://dx.doi.org/10.1021/cr500023c] [PMID: 24895834]

[23]   Lotya, M.; Hernandez, Y.; King, P.J.; Smith, R.J.; Nicolosi, V.; Karlsson, L.S.; Blighe, F.M.; De, S.; Wang, Z.; McGovern, I.T.; Duesberg, G.S.; Coleman, J.N. Liquid phase production of graphene by exfoliation of graphite in surfactant/water solutions. *J. Am. Chem. Soc.,* **2009**, *131*(10), 3611-3620.
[http://dx.doi.org/10.1021/ja807449u] [PMID: 19227978]

[24]   Pumera, M. Electrochemistry of graphene: new horizons for sensing and energy storage. *Chem. Rec.,* **2009**, *9*(4), 211-223.

[http://dx.doi.org/10.1002/tcr.200900008] [PMID: 19739147]

[25]   Bhuyan, M.S.A.; Uddin, M.N.; Islam, M.M.; Bipasha, F.A.; Hossain, S.S. Synthesis of graphene. *Int. Nano Lett.,* **2016**, *6*, 65-83.
[http://dx.doi.org/10.1007/s40089-015-0176-1]

[26]   Choi, W.; Lahiri, I.; Seelaboyina, R.; Kang, Y.S. Synthesis of graphene and its applications: a review. *Crit. Rev. Solid State,* **2010**, *35*, 52-71.
[http://dx.doi.org/10.1080/10408430903505036]

[27]   Novoselov, K.S.; Geim, A.K.; Morozov, S.V.; Jiang, D.; Zhang, Y.; Dubonos, S.V.; Grigorieva, I.V.; Firsov, A.A. Electric field effect in atomically thin carbon films. *Science,* **2004**, *306*(5696), 666-669.
[http://dx.doi.org/10.1126/science.1102896] [PMID: 15499015]

[28]   Liu, J.; Tang, J.; Gooding, J.J. Strategies for chemical modification of graphene and applications of chemically modified graphene. *J. Mater. Chem. A Mater. Energy Sustain.,* **2012**, *22*, 12435-12452.

[29]   Yu, W.; Sisi, L.; Haiyan, Y.; Jie, L. Progress in the functional modification of graphene/graphene oxide: a review. *RSC Advances,* **2020**, *10*, 15328-15345.
[http://dx.doi.org/10.1039/D0RA01068E]

[30]   Li, Y.; Fan, X.; Qi, J.; Ji, J.; Wang, S.; Zhang, G.; Zhang, F. Palladium nanoparticle-graphene hybrids as active catalysts for the Suzuki reaction. *Nano Res.,* **2010**, *3*, 429-437.
[http://dx.doi.org/10.1007/s12274-010-0002-z]

[31]   Li, Y.; Yu, Y.; Wang, J.G.; Song, J.; Li, Q.; Dong, M.; Liu, C.J. CO oxidation over graphene supported palladium catalyst. *Appl. Catal. B,* **2012**, *125*, 189-196.
[http://dx.doi.org/10.1016/j.apcatb.2012.05.023]

[32]   Lange, U.; Hirsch, T.; Mirsky, V.M.; Wolfbeis, O.S. Hydrogen sensor based on a graphene–palladium nanocomposite. *Electrochim. Acta,* **2011**, *56*, 3707-3712.
[http://dx.doi.org/10.1016/j.electacta.2010.10.078]

[33]   Valiollahi, R.; Ojani, R.; Raoof, J.B. Gold nano-cages on graphene support for sodium borohydride electrooxidation. *Electrochim. Acta,* **2016**, *191*, 230-236.
[http://dx.doi.org/10.1016/j.electacta.2016.01.082]

[34]   Guo, X.; Hao, C.; Jin, G.; Zhu, H.Y.; Guo, X.Y. Copper nanoparticles on graphene support: an efficient photocatalyst for coupling of nitroaromatics in visible light. *Angew. Chem. Int. Ed. Engl.,* **2014**, *53*(7), 1973-1977.
[http://dx.doi.org/10.1002/anie.201309482] [PMID: 24505013]

[35]   Song, E.H.; Wen, Z.; Jiang, Q. CO catalytic oxidation on copper-embedded graphene. *J. Phys. Chem. C,* **2011**, *115*, 3678-3683.
[http://dx.doi.org/10.1021/jp108978c]

[36]   Li, A.; Wang, C.; Zhang, H.; Zhao, Z.; Wang, J.; Cheng, M.; Zhao, H.; Wang, J.; Wu, M.; Wang, J. Graphene supported atomic Co/nanocrystalline Co_3O_4 for oxygen evolution reaction. *Electrochim. Acta,* **2018**, *276*, 153-161.
[http://dx.doi.org/10.1016/j.electacta.2018.04.177]

[37]   Bo, Z.; Hu, D.; Kong, J.; Yan, J.; Cen, K. Performance of vertically oriented graphene supported platinum–ruthenium bimetallic catalyst for methanol oxidation. *J. Power Sources,* **2015**, *273*, 530-537.
[http://dx.doi.org/10.1016/j.jpowsour.2014.09.125]

[38]   Wang, R.; Wu, Z.; Chen, C.; Qin, Z.; Zhu, H.; Wang, G.; Wang, H.; Wu, C.; Dong, W.; Fan, W.; Wang, J. Graphene-supported Au-Pd bimetallic nanoparticles with excellent catalytic performance in selective oxidation of methanol to methyl formate. *Chem. Commun. (Camb.),* **2013**, *49*(74), 8250-8252.
[http://dx.doi.org/10.1039/c3cc43948h] [PMID: 23925488]

[39]   Liu, C.H.; Liu, R.H.; Sun, Q.J.; Chang, J.B.; Gao, X.; Liu, Y.; Lee, S.T.; Kang, Z.H.; Wang, S.D. Controlled synthesis and synergistic effects of graphene-supported PdAu bimetallic nanoparticles with

tunable catalytic properties. *Nanoscale,* **2015**, *7*(14), 6356-6362.
[http://dx.doi.org/10.1039/C4NR06855F] [PMID: 25786139]

[40] Saha, S.; Basak, V.; Dasgupta, A.; Ganguly, S.; Banerjee, D.; Kargupta, K. Graphene supported bimetallic GeCoePt nanohybrid catalyst for enhanced and cost effective hydrogen generation. *Int. J. Hydrogen Energy,* **2014**, *39*, 11577.
[http://dx.doi.org/10.1016/j.ijhydene.2014.05.131]

[41] Feng, W.; Yang, L.; Cao, N.; Du, C.; Dai, H.; Luo, W.; Cheng, G. *In situ* facile synthesis of bimetallic CoNi catalyst supported on graphene for hydrolytic dehydrogenation of amine borane. *Int. J. Hydrogen Energy,* **2014**, *39*, 3371-3380.
[http://dx.doi.org/10.1016/j.ijhydene.2013.12.113]

[42] Luo, D.; Zhang, G.; Liu, J.; Sun, X. Evaluation criteria for reduced graphene oxide. *J. Phys. Chem. C,* **2011**, *115*, 11327-11335.
[http://dx.doi.org/10.1021/jp110001y]

[43] He, D.; Cheng, K.; Peng, T.; Sun, X.; Pan, M.; Mu, S. Bifunctional effect of reduced graphene oxides to support active metal nanoparticles for oxygen reduction reaction and stability. *J. Mater. Chem.,* **2012**, *22*, 21298-21304.
[http://dx.doi.org/10.1039/c2jm34290a]

[44] Song, F.Z.; Zhu, Q.L.; Yang, X.; Zhan, W.W.; Pachfule, P.; Tsumori, N.; Xu, Q. Metal–organic framework templated porous carbon-metal oxide/reduced graphene oxide as superior support of bimetallic nanoparticles for efficient hydrogen generation from formic acid. *Adv. Energy Mater.,* **2018**, *8*, 1701416.
[http://dx.doi.org/10.1002/aenm.201701416]

[45] Ali, M.B.; Jo, W.K.; Elhouichet, H.; Boukherroub, R. Reduced graphene oxide as an efficient support for CdS-MoS$_2$ heterostructures for enhanced photocatalytic H$_2$ evolution. *Int. J. Hydrogen Energy,* **2017**, *42*, 16449-16458.
[http://dx.doi.org/10.1016/j.ijhydene.2017.05.225]

[46] Wu, Z.; Song, M.; Zhang, Z.; Wang, J.; Liu, X. Various strategies to tune the electrocatalytic performance of molybdenum phosphide supported on reduced graphene oxide for hydrogen evolution reaction. *J. Colloid Interface Sci.,* **2019**, *536*, 638-645.
[http://dx.doi.org/10.1016/j.jcis.2018.10.068] [PMID: 30391906]

[47] Xu, G.R.; Hui, J.J.; Huang, T.; Chen, Y.; Lee, J.M. Platinum nanocuboids supported on reduced graphene oxide as efficient electrocatalyst for the hydrogen evolution reaction. *J. Power Sources,* **2015**, *285*, 393-399.
[http://dx.doi.org/10.1016/j.jpowsour.2015.03.131]

[48] Mukherjee, A.; Chakrabarty, S.; Su, W.N.; Basu, S. Nanostructured nickel ferrite embedded in reduced graphene oxide for electrocatalytic hydrogen evolution reaction. *Mater. Today Energy,* **2018**, *8*, 118-124.
[http://dx.doi.org/10.1016/j.mtener.2018.03.004]

[49] Ye, B.; Lee, M.; Jeong, B.; Kim, J.; Lee, D.H.; Baik, J.M.; Kim, H.D. Partially reduced graphene oxide as a support of Mn-Ce/TiO$_2$ catalyst for selective catalytic reduction of NOx with NH$_3$. *Catal. Today,* **2019**, *328*, 300-306.
[http://dx.doi.org/10.1016/j.cattod.2018.12.007]

[50] Mahanta, A.; Hussain, N.; Das, M.R.; Thakur, A.J.; Bora, U. Palladium nanoparticles decorated on reduced graphene oxide: An efficient catalyst for ligand-and copper-free Sonogashira reaction at room temperature. *Appl. Organomet. Chem.,* **2017**, *31*, 3679.
[http://dx.doi.org/10.1002/aoc.3679]

[51] Russo, P.; Hu, A.; Compagnini, G. Synthesis, properties and potential applications of porous graphene: a review. *Nano-Micro Lett.,* **2013**, *5*, 260-273.
[http://dx.doi.org/10.1007/BF03353757]

[52]   Zhang, Y.; Wan, Q.; Yang, N. Recent advances of porous graphene: synthesis, functionalization, and electrochemical applications. *Small,* **2019**, *15*(48), e1903780.
[http://dx.doi.org/10.1002/smll.201903780] [PMID: 31663294]

[53]   Guirguis, A.; Maina, J.W.; Zhang, X.; Henderson, L.C.; Kong, L.; Shon, H.; Dumée, L.F. Applications of nano-porous graphene materials–critical review on performance and challenges. *Mater. Horiz.,* **2020**, *7*, 1218-1245.
[http://dx.doi.org/10.1039/C9MH01570A]

[54]   Qiu, H.; Dong, X.; Sana, B.; Peng, T.; Paramelle, D.; Chen, P.; Lim, S. Ferritin-templated synthesis and self-assembly of Pt nanoparticles on a monolithic porous graphene network for electrocatalysis in fuel cells. *ACS Appl. Mater. Interfaces,* **2013**, *5*(3), 782-787.
[http://dx.doi.org/10.1021/am3022366] [PMID: 23331257]

[55]   Xia, W.Y.; Li, N.; Li, Q.Y.; Ye, K.H.; Xu, C.W. Au-NiCo$_2$O$_4$ supported on three-dimensional hierarchical porous graphene-like material for highly effective oxygen evolution reaction. *Sci. Rep.,* **2016**, *6*, 23398.
[http://dx.doi.org/10.1038/srep23398] [PMID: 26996816]

[56]   Guo, B.; Fang, L.; Zhang, B.; Gong, J.R. Graphene doping: a review. *Insciences J.,* **2011**, *1*, 80-89.
[http://dx.doi.org/10.5640/insc.010280]

[57]   Putri, L.K.; Ong, W.J.; Chang, W.S.; Chai, S.P. Heteroatom doped graphene in photocatalysis: a review. *Appl. Surf. Sci.,* **2015**, *358*, 2-14.
[http://dx.doi.org/10.1016/j.apsusc.2015.08.177]

[58]   Wang, H.; Maiyalagan, T.; Wang, X. Review on recent progress in nitrogen-doped graphene: synthesis, characterization, and its potential applications. *ACS Catal.,* **2012**, *2*, 781-794.
[http://dx.doi.org/10.1021/cs200652y]

[59]   Ewels, C.P.; Glerup, M. Nitrogen doping in carbon nanotubes. *J. Nanosci. Nanotechnol.,* **2005**, *5*(9), 1345-1363.
[http://dx.doi.org/10.1166/jnn.2005.304] [PMID: 16193950]

[60]   Groves, M.N.; Chan, A.S.W.; Malardier-Jugroot, C.; Jugroot, M. Improving platinum catalyst binding energy to graphene through nitrogen doping. *Chem. Phys. Lett.,* **2009**, *481*, 214-219.
[http://dx.doi.org/10.1016/j.cplett.2009.09.074]

[61]   Xiong, B.; Zhou, Y.; Zhao, Y.; Wang, J.; Chen, X.; O'Hayre, R.; Shao, Z. The use of nitrogen-doped graphene supporting Pt nanoparticles as a catalyst for methanol electrocatalytic oxidation. *Carbon,* **2013**, *52*, 181-192.
[http://dx.doi.org/10.1016/j.carbon.2012.09.019]

[62]   Gao, M.R.; Cao, X.; Gao, Q.; Xu, Y.F.; Zheng, Y.R.; Jiang, J.; Yu, S.H. Nitrogen-doped graphene supported CoSe$_2$ nanobelt composite catalyst for efficient water oxidation. *ACS Nano,* **2014**, *8*(4), 3970-3978.
[http://dx.doi.org/10.1021/nn500880v] [PMID: 24649855]

[63]   Xu, Y.; Tu, W.; Zhang, B.; Yin, S.; Huang, Y.; Kraft, M.; Xu, R. Nickel Nanoparticles Encapsulated in Few-Layer Nitrogen-Doped Graphene Derived from Metal-Organic Frameworks as Efficient Bifunctional Electrocatalysts for Overall Water Splitting. *Adv. Mater.,* **2017**, *29*(11), 1605957.
[http://dx.doi.org/10.1002/adma.201605957] [PMID: 28102612]

[64]   Jayaramulu, K.; Masa, J.; Tomanec, O.; Peeters, D.; Ranc, V.; Schneemann, A.; Zboril, R.; Schuhmann, W.; Fischer, R.A. Nanoporous nitrogen-doped graphene oxide/nickel sulfide composite sheets derived from a metal-organic framework as an efficient electrocatalyst for hydrogen and oxygen evolution. *Adv. Funct. Mater.,* **2017**, *27*, 1700451.
[http://dx.doi.org/10.1002/adfm.201700451]

[65]   Jafri, R.I.; Rajalakshmi, N.; Ramaprabhu, S. Nitrogen doped graphene nanoplatelets as catalyst support for oxygen reduction reaction in proton exchange membrane fuel cell. *J. Mater. Chem.,* **2010**,

20, 7114-7117.
[http://dx.doi.org/10.1039/c0jm00467g]

[66]   Wu, Z.S.; Yang, S.; Sun, Y.; Parvez, K.; Feng, X.; Müllen, K. 3D nitrogen-doped graphene aerogel-supported Fe$_3$O$_4$ nanoparticles as efficient electrocatalysts for the oxygen reduction reaction. *J. Am. Chem. Soc.,* **2012**, *134*(22), 9082-9085.
[http://dx.doi.org/10.1021/ja3030565] [PMID: 22624986]

[67]   Sun, Y.; Du, C.; An, M.; Du, L.; Tan, Q.; Liu, C.; Gao, Y.; Yin, G. Boron-doped graphene as promising support for platinum catalyst with superior activity towards the methanol electrooxidation reaction. *J. Power Sources,* **2015**, *300*, 245-253.
[http://dx.doi.org/10.1016/j.jpowsour.2015.09.046]

[68]   Damte, J.Y.; Lyu, S.L.; Leggesse, E.G.; Jiang, J.C. Methanol decomposition reactions over a boron-doped graphene supported Ru-Pt catalyst. *Phys. Chem. Chem. Phys.,* **2018**, *20*(14), 9355-9363.
[http://dx.doi.org/10.1039/C7CP07618E] [PMID: 29564450]

[69]   Joshi, P.; Huang, H.H.; Yadav, R.; Hara, M.; Yoshimura, M. Boron-doped graphene as electrocatalytic support for iridium oxide for oxygen evolution reaction. *Catal. Sci. Technol.,* **2020**, *10*, 6599-6610.
[http://dx.doi.org/10.1039/D0CY00919A]

[70]   Hu, M.; Yao, Z.; Li, L.; Tsou, Y.H.; Kuang, L.; Xu, X.; Zhang, W.; Wang, X. Boron-doped graphene nanosheet-supported Pt: a highly active and selective catalyst for low temperature H$_2$-SCR. *Nanoscale,* **2018**, *10*(21), 10203-10212.
[http://dx.doi.org/10.1039/C8NR01807C] [PMID: 29786726]

[71]   Yang, Y.; Castano, C.E.; Gupton, B.F.; Reber, A.C.; Khanna, S.N. A fundamental analysis of enhanced cross-coupling catalytic activity for palladium clusters on graphene supports. *Nanoscale,* **2016**, *8*(47), 19564-19572.
[http://dx.doi.org/10.1039/C6NR06793J] [PMID: 27833943]

[72]   Yang, Y.; Reber, A.C.; Gilliland, S.E., III; Castano, C.E.; Gupton, B.F.; Khanna, S.N. More than just a support: Graphene as a solid-state ligand for palladium-catalyzed cross-coupling reactions. *J. Catal.,* **2018**, *360*, 20-26.
[http://dx.doi.org/10.1016/j.jcat.2018.01.027]

CHAPTER 5

Graphene-Based Nano-materials as Catalyst for the Synthesis of Medicinally Implanted Scaffold 1,2,3-triazoles

Jasmin Sultana[1] and **Diganta Sarma**[1,*]

[1] *Department. of Chemistry, Dibrugarh University, Dibrugarh-786004, Assam, India*

Abstract: Graphene-based nano-materials have received great attention in energy and environmental applications. Their unique physicochemical assets, like high electron mobility, high thermal stability, high surface area, chemical flexibility, and mechanical solidity, make them highly flexible materials for versatile applications. In this chapter, the application of graphene-based nano-materials for the synthesis of 1,2,3-triazole scaffolds is discussed in detail. As graphene possesses high carrier mobility, it enhances the copper-catalyzed 1,3-dipolar cycloaddition of azides and alkynes, leading to the formation of 1,4-disubstituted 1,2,3-triazoles.

Keywords: 1,2,3-triazoles, Cycloaddition reactions, Graphene, Graphene oxide, Heterogeneity, Nanoparticles, Reduced graphene oxide.

INTRODUCTION

The most important nitrogen-containing heterocyclic motif, 1,2,3-triazole, occupies a wide area of the research community due to its diversified applications in pharmaceuticals, biological, organic chemistry, organometallic chemistry, material, agrochemical, and industrial fields [1 - 6]. In comparison to the other three adjacent nitrogen-containing heterocyclic scaffolds, 1,2,3-triazoles are stable under typical physiological conditions. Moreover, triazole moieties are highly susceptible to H-bonding and can easily interact with biomolecules. 1,2,3 triazoles are equipped with a broad spectrum of pharmacological activities [7], such as anti-tuberculosis, anticancer, anti-HIV, antibacterial, and antifungal activities. Some typical bioactive 1,2,3-triazoles are represented in Fig. (**1**). Therefore, the development of advanced synthetic methodologies for facile construction of 1,2,3-triazoles is in high demand.

---
* **Corresponding author Diganta Sarma:** Department of Chemistry, Dibrugarh University, Dibrugarh-786004, Assam, India; E-mail: dsarma22@dibru.ac.in

Manorama Singh, Vijai K. Rai and Ankita Rai (Eds)

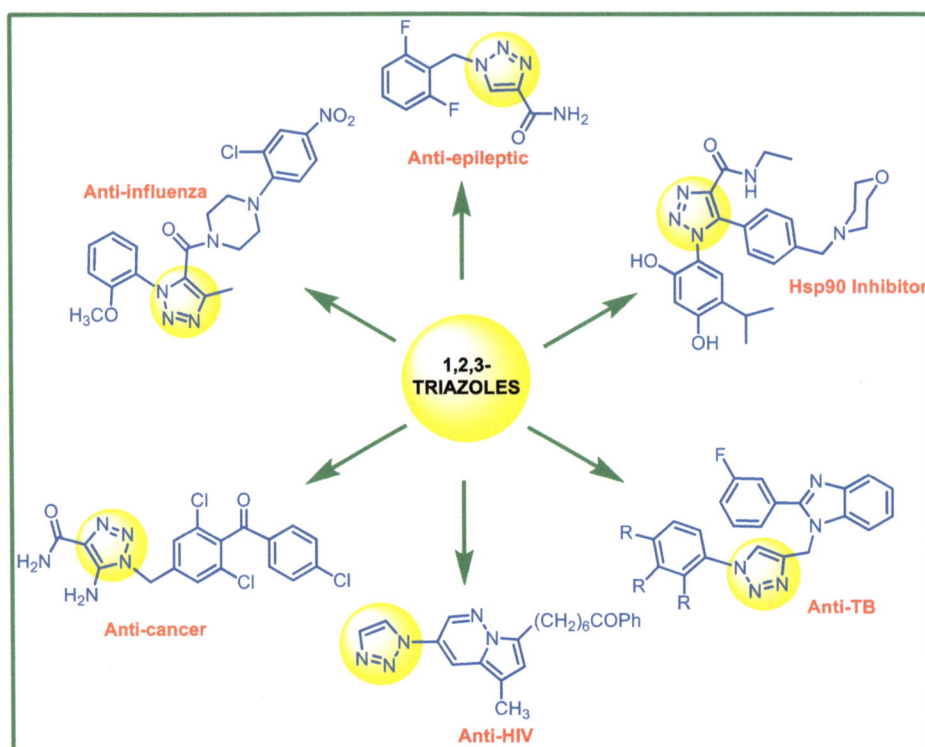

Fig. (1). Bioactivity of 1,2,3-triazoles.

1,2,3-triazoles have two regioisomers, I,4-disubstituted and 1,5-disubstituted 1,2,3-triazoles. 1,4,5-trisubstituted-1,2,3-triazoles have also gained remarkable attention in the past few years. R. Huisgen was the first who introduced the 1,3-dipolar cycloaddition of azides and alkynes, producing both the regioisomers [8]. Later in 2002, the non-selectivity and high-temperature requirement of this conventional method were overcome by the independent and pioneering work of Sharpless *et al.* and Mendal *et al.* [9, 10]. The use of copper sulfate as a catalyst and sodium ascorbate as a reducing agent provided the 1,4-regioisomer under copper-catalyzed azide-alkyne cycloaddition reaction (CuAAC) at room temperature (Scheme **1**). This work on CuAAC can be considered as a landmark for efficient formation of 1,4-disubstituted 1,2,3-triazoles as a sole regioisomer, drawing the attention of the scientific community towards it. Subsequently, numerous successful works have been done using different copper sources as a homogeneous catalytic system [11 - 15]. The homogeneous catalytic systems are associated with disadvantages such as non-reusability, use of expensive ligands and bases, formation of metal complexes, and metal contamination in the final products. Thus, from an environmental concern, heterogeneous catalysis has gained much more importance over homogeneous copper catalysts. Heterogeneous catalysis also eliminates the possibility of bistriazole and diacetylene formations and reduces the effective copper loading for CuAAC reactions. Recently, copper immobilized on different supports, such as zeolites, chitosan, clay, cellulose, charcoal, hydrotalcite, *etc.*, are used as heterogeneous catalysts for the synthesis of fluorescent bioactive 1,4-disubstituted 1,2,3-triazoles [16 - 23]. As the design and synthesis of catalysts with superior catalytic performance have become a crucial area of research, researchers are continuously trying to develop more and more advanced methodologies for 1,2,3-triazole synthesis.

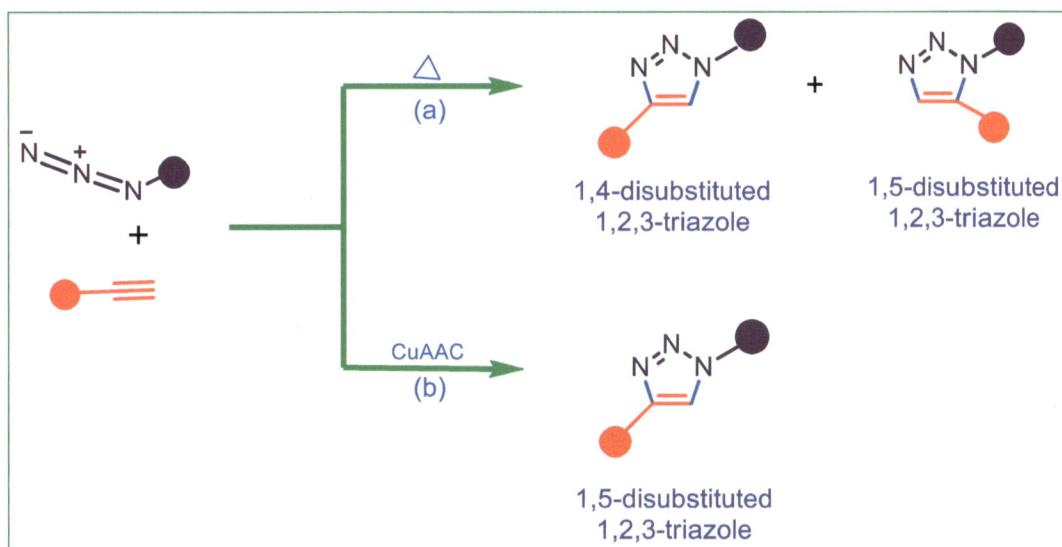

Scheme (1). Huisgen and Sharpless' AAC reaction.

Graphene, a carbon-based support with 2D aromaticity and honeycomb-like structure, is gaining much attention from the scientific community [24]. Graphene is associated with certain unique properties, like high optical transmittance, high thermal conductivity, high Young's modulus, huge surface area, and high mobility of charge carriers. The collective effects of these properties are responsible for the fabrication of graphene-based nanocomposites, which have huge applications in different fields, like nano-electronics, supercapacitors, batteries, fuel cells, solar cells, electrocatalysis, nanocomposites, and sensing [25]. The high specific surface area makes graphene a potential material for catalytic applications.

Over the past few decades, nanotechnology has been gaining much attention in the field of synthetic organic chemistry. The higher surface-to-volume ratio makes nanoparticles superior over the bulk catalysts. However, the process of agglomeration is the major drawback associated with these bare nanoparticles, and this problem can be overcome by the use of a stabilizer. Chemists are trying to use graphene as internal stabilizer and also as a support for immobilization of various metal and metal oxide nanoparticles such as Ag [26], Pt [27], Au [28], Pd [29], Co [30], Sn [31], MnO_2, TiO_2 [32], SnO_2 and ZnO [33]. Thus, controlled and uniform loading of different nanocatalysts of interest, high surface to volume ratio, reusability, and the stability factor enhance the application of graphene-based nanomaterials in the field of catalysis.

The carbon skeleton, graphene, can accelerate the chemical conversions by making the aromatic reactants more accessible to the active sites of active metal species [34]. As part of continuing interest in the perfection of feasibility and eco-friendliness for synthesizing 1,4-disubstituted 1,2,3-triazoles, chemists are trying to use graphene-based nanomaterials as catalysts for azide-alkyne cycloaddition reactions. Direct azide-alkyne cycloaddition and one-pot three-component cycloadditions are two major routes in which these nanomaterials are found to have a lot of applications.

GRAPHENE-BASED NANOMATERIALS IN DIRECT AZIDE-ALKYNE CYCLOADDITION REACTIONS

Graphene oxide (GO), a modified form of graphene, can be conveniently prepared on a large scale by controlled chemical reactions, and it acts as a host to install metal nanoparticles by preventing their agglomeration. The presence of different oxygen functionalities, such as hydroxyl (-OH), epoxy (C-O-C), carboxyl (-COOH), increases the hydrophilicity of the GO surfaces. This hydrophilic nature, large surface area, and layered structure enhance the deposition of different metals and metal oxides onto the GO surfaces [35, 36]. Taking these advantages, A. S. Nia and coworkers have used copper nanoparticles supported on graphene for CuAAC reactions for the first time [37]. The catalyst, TRGO/Cu (I), used in the reaction was synthesized by the continuous stirring of GO with copper acetate in water, followed by the reduction in Ar at 600°C (Scheme **2**).

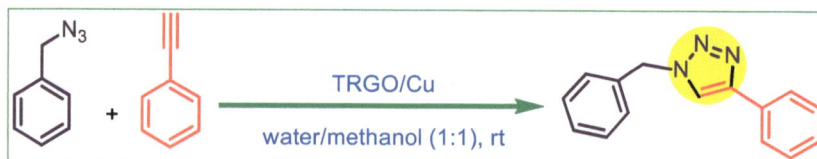

Scheme 2. TRGO/Cu catalyzed AAC reaction.

The Cu$_2$O-GO nanocomposite was successfully used in the reaction between benzyl azide and different aromatic and aliphatic azides leading to high yields of the desired products. The authors have analyzed the recyclability and reusability of the catalyst in the model CuAAC reaction. According to authors, the 30% reduction of the catalytic performance after three successive cycles could be due to the process of aggregation of the copper nanoparticles. The bulk analysis of the catalyst reveals its excellency for CuAAC reactions, as proven by melt-rheology. In 2016, V. H. Reddy's group developed a green catalytic methodology for 1,2,3-triazole synthesis using copper oxide nanocomposite supported on GO (Scheme **3**) [38].

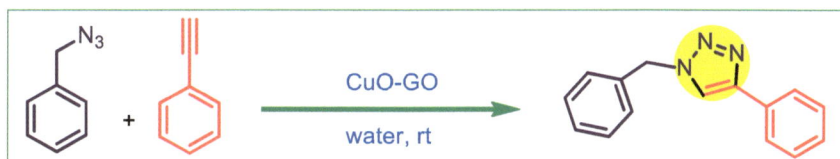

Scheme 3. CuO-GO catalyzed AAC reaction.

The uniform distribution and well deposition of CuO nanoparticles over the surface of GO was confirmed by various analytical techniques, like TEM and XRD. Due to the combined advantages of GO and CuO, the catalyst was effectively applied to CuAAC reactions using water as a green solvent. Ligand-free environment, low catalyst loading, aerobic conditions, and room temperature requirement are some useful advantages of this strategy. A variety of azides and alkynes with different functionality are tolerated under these conditions. The high catalytic performance is due to the dispersity of the catalyst in water and easy interaction with organic reagents. "Breslow effect" is the influence of water as solvent. It enhances the catalytic activity [39]. The recyclability and reusability analysis of the catalyst shows the retention of its catalytic activity for up to five consecutive cycles. The partial leaching of active copper species after five consecutive runs leads to the lowering of its catalytic activity. Easy isolation of the triazole products directly from the reaction flask, eliminating the need for column chromatography, is the major advantage of this methodology.

Y. Jain and coworkers have fabricated a magnetically separable nanocatalyst, GO-Fe$_3$O$_4$@CuO, and successfully applied it for the synthesis of sugar-coumarin-based substituted 1,2,3-triazoles under ultrasonication [40]. TEM (Transmission Electron Microscopy) and HRTEM (High-Resolution Transmission Electron Microscopy) analysis showed the dispersion of Fe$_3$O$_4$ and CuO particles over multilayer GO sheets. The presence of copper in Cu (II) state is confirmed by XPS analysis.

The catalytic performance was accessed by using sugar azide and coumarin alkynes as substrates (Scheme **4**).

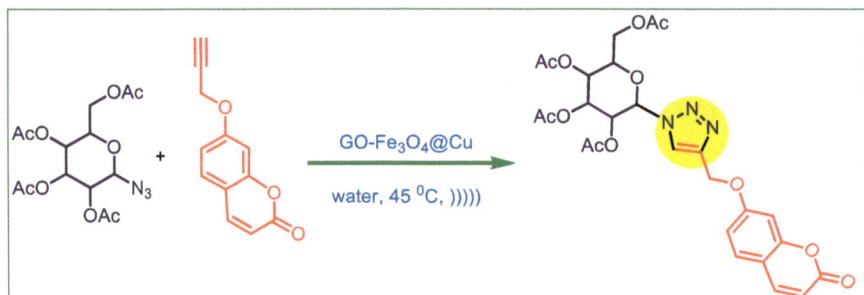

Scheme 4. GO-Fe$_3$O$_4$@Cu catalyzed AAC reaction.

The noticeable features of this strategy are simple reaction conditions, easy work-up procedure, high yields of products, and high stability of the catalyst. In addition to excellent activity and selectivity, easy removal of the catalyst from the reaction mixture using a simple permanent magnet is also possible. The reusability of the catalyst up to eight consecutive runs without dropping its activity makes it an interesting protocol for triazole synthesis. The cytotoxicity analysis of the representative compounds against PC-12 cell lines displayed that compounds having trifluoromethyl groups give the best result with IC$_{50}$ (8.3µg/mL) compared to the standard drug cisplatin (5.8 µg/mL). Another efficient, green, simple, and scale-up protocol for synthesizing biologically significant substituted 1,2,3-triazoles using coumarin azide derivatives and terminal alkynes was developed by Y. Jain's group in 2019 (Scheme **5**) [41]. CuO/GO nanocomposite under ultrasonic irradiation at 45 ^{0}C catalyzes the CuAAC reactions in water, providing various triazole derivatives in good yields. TEM images of the CuO/GO nanocomposite revealed the presence of spindle-shaped CuO NPs embedded on GO sheets. These CuO nanoparticles are constituted by some smaller nanorods. From XPS analysis, binding energies of 934.5 eV and 954.20 eV for Cu 2P$_{3/2}$ and Cu 2p$_{1/2,}$ respectively, suggested the presence of Cu^{2+} in the catalyst. The following synergic effects enhance the catalytic activity:

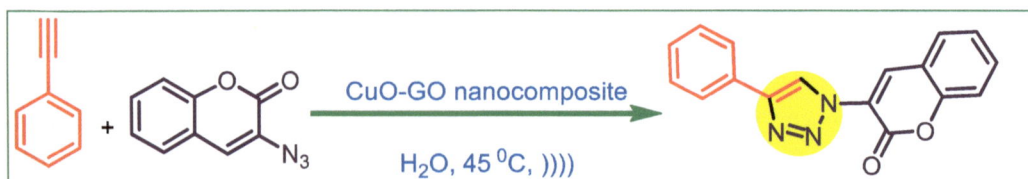

Scheme 5. CuO-GO nanocomposite in triazole synthesis.

a. High absorption ability of GO towards reactants through π-π interactions, providing more places for interactions and thus accelerate the reactions.
b. High concentrations of reactants near the active CuO NPs.
c. The "Breslow effect."

The most useful aspects of this protocol are that it is devoid of byproducts and its recyclability and reusability, which is up to the ninth cycle. AAS (Atomic Absorption Spectroscopy) revealed the negligible difference between the reused catalyst and the fresh one, indicating the low leaching of CuO NPs into the reaction mixture.

GRAPHENE-BASED NANOMATERIALS IN ONE-POT AZIDE-ALKYNE CYCLOADDITION REACTIONS

The organic azides required for azide-alkyne cycloaddition (AAC) reactions are very harmful to human health. Therefore, to avoid its direct handling in the laboratory, the *in situ* prepared azides have gained much importance in synthetic chemistry. For this, a one-pot azide-alkyne cycloaddition reaction is used in which diazotization of anilines or amines followed by the addition of sodium azide gives the corresponding organic azides. The direct displacement reaction between alkyl or aromatic halides with sodium azide is also useful in this regard. Thus, the one-pot cycloaddition of *in situ* generated azides with alkynes under milder reaction conditions is an attractive research area.

With the intention of applying graphene-based NPs in one-pot AAC reactions, a new, safe, robust, and magnetically recoverable heterogeneous catalyst, graphene-based material with γ-Fe_3O_4 nanoparticles was prepared and characterized by TEM, XRD, and Raman spectroscopy by N. Salam *et al.* in 2013 [42]. The catalyst, with its high activity and selectivity properties, accelerates the reaction between alkynes and *in situ* generated azides leading to different 1,4-disubstituted 1,2,3-triazole derivatives (Scheme **6**). The aromatic halides with increased electronic effects increase the product yields, whereas electron deficient aromatic halides give negative results. The Prussian blue test confirmed the oxidation state of iron in the graphene- γ-Fe_3O_4 nanocomposite as +3. According to the authors, the improvement of catalytic activity can be credited to the following reasons:

a. High migration efficiency of electrons due to the conducting property of graphene [43].
b. Facile loss of catalytic activity was obstructed by preventing aggregation of graphene and γ-Fe_3O_4 NPs. The high surface area of the catalyst provides more adsorption of reactants.
c. Graphene acts both as a catalyst and support for the reaction.

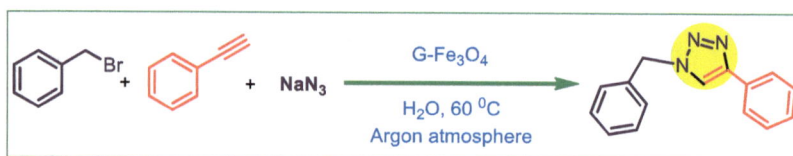

Scheme 6. G-Fe$_3$O$_4$ catalyzed MCRs for triazole synthesis.

The superparamagnetic nature of γ-Fe$_3$O$_4$ NPs leads to easy separation of reacted catalyst from the reaction mixture using a bar magnet. The leaching test indicates the heterogeneous nature of the catalyst, and its reusability up to five runs makes it an industrially important methodology for organic synthesis.

Due to their unique chemical properties, silver and its salts have been used as catalysts and promoters in organic synthesis. Their excellent catalytic activity towards cycloaddition, cycloisomerization, coupling, nitrene transfer rearrangement, and sigmatropic rearrangement motivated the scientists to introduce silver catalyzed azide-alkyne cycloaddition (AgAAC) in one-pot methodology (Scheme 7) [44]. Ag-nanoparticles immobilized on different supports, *viz.* graphene, graphene oxide, and carbon nanotubes, are used for different antibacterial and electrochemical applications [45]. However, their applications in 1,2,3-triazole synthesis were first introduced by N. Salam and coworkers in 2014 [46]. They have developed a graphene-based composite with silver nanoparticles (Ag-G) using a straightforward chemical route characterized by XRD, UV-Visible spectroscopy, Raman spectroscopy, and TEM. The uniform dispersion of Ag NPs over the graphene sheets and their composite nature can be clearly seen from the low and high-resolution TEM images of the catalyst. By considering Ag-G as a catalyst of choice, aniline was diazotized under appropriate conditions and treated with phenylacetylene in water at room temperature. The reaction of anilines having a different electronic environment with phenylacetylene shows a high reaction rate with aryl, benzyl, and alkyl azides at room temperature or 40 °C. The explanation of high catalytic activity for Cu@GO NPs is also applicable for Ag@GO NPs, *i.e.*, the synergistic effect of GO. This synergistic effect facilitates highly efficient contact between GO and reactant molecules and uptaking of more electrons by reactant molecules from GO sheets. These heterogeneous catalysts are highly stable. No silver leaching, prevention of agglomeration, and recyclability up to multiple times increase their useful applications in green organic synthesis.

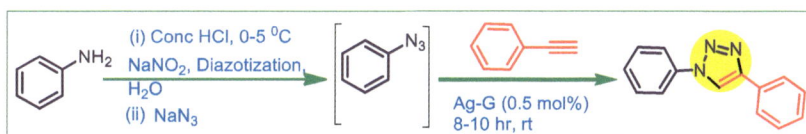

Scheme 7. One-pot methodology for AgAAC reaction.

Considering the intrinsic use of graphene and cuprous oxide (Cu_2O) in different areas, like catalysis, medicinal chemistry, biosensors, gas sensors, transistors, and lithium ion batteries, graphene-Cu_2O composites are synthesized. They are expected to show novel properties, like superior electrochemical activity and better charge-discharge capacity [47]. Thus, I. Roy and coworkers have synthesized a reduced graphene oxide/cuprous oxide (RGO/Cu_2O) nanocomposite by an *in situ* reduction methodology using lactulose in aqueous media at high temperature and pressure [48]. Lactulose, being a disaccharide of fructose and galactose, acts as a reducing-cum-stabilizing agent for the synthesis of the catalyst under environmentally benign conditions. The formation of Cu_2O nanoparticles and reduction of GO take place in a one-pot reaction without using any toxic reagents.

The smaller particle size of Cu_2O in the RGO/Cu_2O nanocomposite than the pure Cu_2O nanoparticles was confirmed from their respective TEM images. The synthesis of RGO/Cu_2O nanocomposite was confirmed by energy-dispersive X-ray spectroscopy, X-ray diffraction, UV-visible spectroscopy, field emission scanning electron microscopy, Fourier-transform infrared spectroscopy, and transmission electron microscopy. The catalyst RGO/Cu_2O nanocomposite was successfully applied for cycloaddition between *in situ* prepared azides and alkynes, affording 1,2,3-triazoles in excellent yields (Scheme **8**). The authors have analyzed three different benzyl bromide derivatives as an initial precursor of organic azide. The reactions proceeded smoothly with 91-95% isolated yields of products. Recyclability of the catalyst for up to six consecutive cycles showed its unaltered catalytic activity.

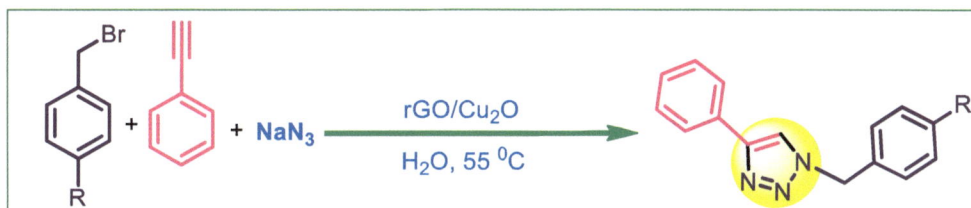

Scheme 8. rGO/Cu_2O catalyzed MCRs for triazole synthesis.

A new heterogeneous copper catalyst, GO/poly (vinyl imidazole) nanocomposite, was synthesized by A. Pourjavadi's group in 2015 [49]. GO was entrapped into the polymer support, and then copper was adsorbed onto the nanocomposite matrix (GO/Pim/Cu). The cross-linked polymer having multiple layers enhanced the loading of copper metal onto the catalyst. According to the authors, most of the Cu^{2+} ions were coordinated with two different imidazole rings of the polymer chains by complexation. It was also observed that Cu^{2+} ions were not embedded into the internal layers of the catalyst, and all the imidazole rings of the polymer chains were not coordinated to the metal ion.

The prepared catalyst, GO/Pim/Cu was then successfully applied for a one-pot three-component cycloaddition of benzyl bromide, sodium azide, and phenylacetylene to synthesize different triazole derivatives (Scheme 9). From an environmental point of view, 1 mol% of the catalyst at 50 ^{0}C in aqueous media was chosen as the optimized condition for this cycloaddition reaction. Higher activity of GO/Pim/Cu over Pim/Cu suggests the requirement of porous GO sheets for the reaction. Both aromatic and aliphatic alkynes readily undergo 1,3-dipolar cycloaddition with both alkyl and benzyl azides leading well to excellent yields of 1,2,3-triazole products under mild reaction conditions.

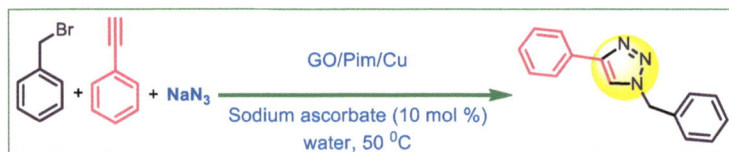

Scheme 9. GO/Pim/Cu catalyzed MCRs for triazole synthesis.

For supported metal catalysts, contamination of products with leached metals is a serious issue. Metal leaching leads to a decrease in turn-over frequency (TOF) and involves costly clean-up and time-consuming steps. However, the positive heterogeneity test of the catalyst nullifies these problems. High product purity, operational simplicity, recyclability up to eight cycles, and avoidance of hazardous organic solvents make it a useful and attractive scheme for large-scale synthesis in organic chemistry.

In the past few decades, researchers have reported different metals, including Ag/Pd [50], Pd/Cu [51], and Pd/Fe$_3$O$_4$ [52], supported on rGO (reduced graphene oxide) as heterogeneous catalysts. These catalysts are relatively more expensive. On the other hand, copper nanoparticles are also unstable in air and easily oxidized to their corresponding oxide form. Therefore, in 2018, Z. Li *et al*. have prepared an rGO supported Cu-Cu$_2$O nanocomposite (Cu-Cu$_2$O@rGO) through a multicomponent one-pot reflux methodology. The stability of Cu-Cu$_2$O nanoparticles over the rGO matrix is due to their highly delocalized electrons [53]. The non-toxic and inexpensive reagent, glucose, reduces GO and copper sulfate, whereas polyvinylpyrrolidone as a green surfactant effectively controls the size of metal nanocomposites.

Cu-Cu$_2$O@rGO catalyst was characterized by different analytical techniques, like X-ray photoelectron, transmission electron microscopy, infrared and Raman spectroscopy, X-ray diffraction, *etc*. The Cu–Cu$_2$O nanoparticles -exhibited fine dispersion on rGO with an average size of about 15 nm. Application of this heterogeneous catalyst to tandem reactions of alkynes, sodium azide, and halides provided the respective 1,4-disubstituted 1,2,3-triaoles in excellent yields (Scheme **10**) [54]. This catalyst has certain advantages in tandem reactions, like simple methods for catalyst preparation, high yields of products, recyclability *etc*. Successful use of 5 mol % of the catalyst at room temperature for two hours with very little leaching of metal shows its excellent catalytic performance in the one-pot cycloaddition strategy.

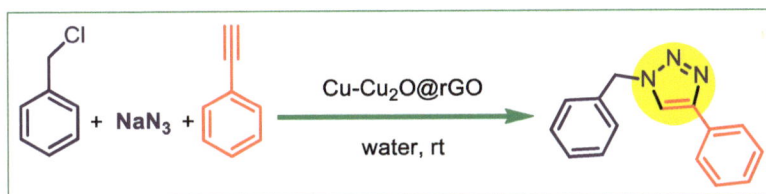

Scheme 10. Cu-Cu$_2$O@rGO catalyzed MCRs for triazole synthesis.

Chitosan is a deacetylated product of chitin and a major component of the exoskeleton of shrimp and other crustaceans [55]. The simultaneous presence of hydroxyl and amino groups makes it the best bio adsorbent of metals from water and wastewater. Magnetic chitosan complexes are also useful for the removal of dyes and toxic metals from aqueous media [56]. Realizing the importance of magnetic heterocatalysts, the good adsorption capacity of Cu^{2+} ions and chitosan, M. Mahdavinasab and coworkers have developed a magnetic chitosan-based nanocomposite, MnFe$_2$O$_4$@GO@Chitosan/Cu [57]. The analytical techniques XRD, SEM, TEM, FT-IR, TGA, and VSM revealed the thermal stability, structure, and magnetic properties of the catalyst.

Its application as a heterogeneous catalyst for condensation of terminal alkynes, primary halides, and sodium azide in the presence of sodium ascorbate leads to high yields of products in short reaction times (Scheme **11**). Avoiding hazardous solvents is an important perspective of the reaction. Analysis of substituent effect shows the clean and easy formation of triazole moieties in excellent yields within 1-2 hours. The hot filtration test confirms the heterogeneity of the catalyst by showing negligible leaching of the metal during the reaction. The catalyst is magnetically detachable and is reused many times with significant catalytic activity.

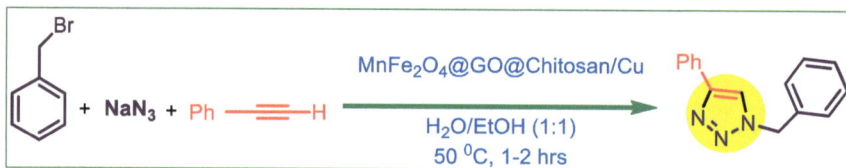

Scheme 11. MnFe$_2$O$_4$@GO@CS/Cu-catalyzed MCRs for triazole synthesis.

As copper is cytotoxic in nature; therefore, the use of reduced quantities of copper@GO is an appealing approach. On the other hand, the utilization of light for chemical transformations is advantageous from an environmental point of view. Therefore, photochemistry, in addition to heterogeneous catalysis, becomes an important aspect of green chemistry as it shows decreased adverse effects on the environment. Regarding these factors, M. M. Aghayan and coworkers have synthesized a photocatalyst, Cu$_2$O/reduced graphene oxide/TiO$_2$ (Cu$_2$O/rGO/TiO$_2$) from TiO$_2$, graphene oxide, and CuCl$_2$ under ultrasonic irradiation [58]. The nanocomposite was fully characterized by different analytical techniques.

The Cu$_2$O/rGO/TiO$_2$ photocatalyst was efficiently used for the synthesis of substituted 1,2,3-triazoles under multicomponent reactions of epoxides or halides with terminal alkynes and NaN$_3$ in the presence of a base. The one-pot multicomponent reaction of organic halides with alkynes and NaN$_3$ requires Cu$_2$O/rGO (0.5%)/TiO$_2$ as nanocatalyst, DMF: H$_2$O (3:1) as a solvent, triethylamine as a base, and visible radiation of 450 nm (Scheme **12**). The presence of base is observed to be essential for this type of condensation reaction as it is expected to stabilize and protect the copper ion in a +1 oxidation state in the nanocomposite. It also acts as a hole-scavenger in this cycloaddition reaction. A range of benzyl halides and alkynes with different functionality were observed to be tolerated under these conditions giving sufficient yields of respective products.

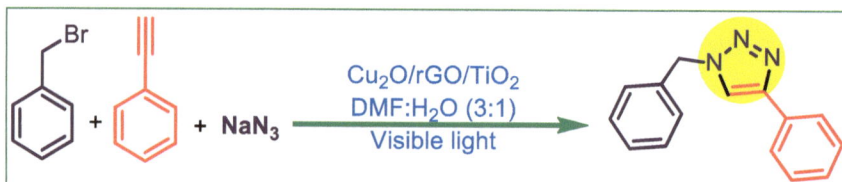

Scheme 12. Cu$_2$O/rGO/TiO2 catalyzed AAC reaction.

The same optimized condition is also applicable for the synthesis of β-hydroxy triazoles through the reaction of epoxides with alkynes and NaN_3 in the presence of triethylamine (Scheme **13**). Alkyl and aryl bearing groups and cyclic epoxides are observed to undergo this type of cycloaddition reaction, providing good to excellent yields of respective triazoles. The observed regioselectivity of this reaction is highly remarkable. When the substituent R of the epoxide ring is an alkyl group, the nucleophilic attack will occur on the less substituted carbon of the epoxide ring. Alternatively, the nucleophilic attack will occur at the more hindered carbon of the epoxide ring when R is an aryl group.

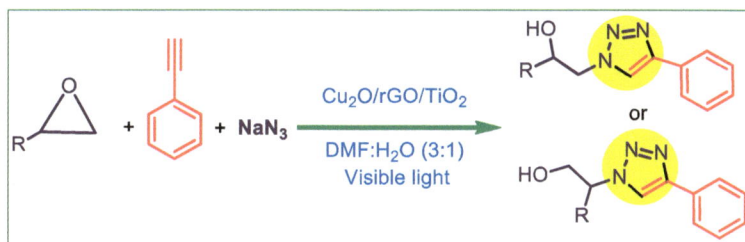

Scheme 13. One-pot methodology for triazole synthesis using $Cu_2O/rGO/TiO_2$.

F. Rafiee and coworkers have synthesized magnetic graphene oxide *via* chemical oxidation, irradiation with ultrasonic waves, and magnetization process [59]. The cysteine incorporated magnetic graphene oxide was then used as efficient support for copper immobilization to prepare $Cu^{I/II}$@Cys-MGO□nanocomposite. XPS analysis confirmed the presence of copper in both +1 and +2 oxidation states. Some Cu^{2+} ions are reduced to Cu^{1+} ions due to incomplete oxidation of cysteine to cystine.

The $Cu^{I/II}$@Cys-MGO nanocatalyst was then efficiently applied for two consecutive reactions: (a) synthesis of aryl azides from aryl boronic acids and (b) 1,4-diaryl triazole synthesis through one-pot reaction mechanisms. The coupling reaction of 2-methoxy-phenyl boronic acid, NaN_3, and phenylacetylene shows the best result at 60 0C when a 1:1 mixture of water and ethanol was used (Scheme **14**). Various electronically different boronic acid derivatives and alkynes have provided good yields of the respective 1,2,3-triazoles. High product yields, low amount of copper, short reaction time, no copper contamination, mild reaction pathway, and reusability make the system more efficient and cost-effective.

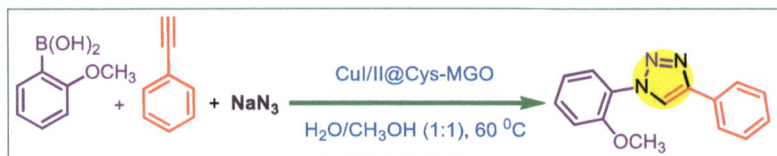

Scheme 14. One-pot methodology for triazole synthesis using CuI/II@Cys-MGO.

CONCLUSION

Graphene is a rising star material and is gaining more attention from researchers after it was first reported by Novoselov *et al*. in 2004. Certainly, graphene-based materials have exhibited great potential in environmental remediation. The high chemical and thermal stabilities promote graphene to be suitable support material for nanocatalysts. Graphene (G), graphene oxide (GO), and reduced graphene oxide (rGO) act as excellent support materials for the immobilization of different metal and metal oxide nanoparticles. Their successful applications in the field of direct azide-alkyne cycloadditions and one-pot multi-component reactions explore their catalytic activity in the area of synthetic organic chemistry. Easy recyclability, reusability up to several cycles, and low or no leaching of the metal into the reaction mixture prove the heterogeneous nature of these supported metal catalysts. These catalytic systems provide a major contribution to the field of development of advanced and efficient strategies for 1,4-disubstituted 1,2,3-triazole synthesis.

CONSENT FOR PUBLICATION

Not applicable.

CONFLICT OF INTEREST

The author declares no conflict of interest, financial or otherwise.

ACKNOWLEDGEMENTS

We are thankful to DST, New Delhi, India, for the research grant [No. EMR/2016/002345]. We gratefully acknowledge the Department of Science and Technology for financial assistance under the DST-FIST program and UGC, New Delhi, for the Special Assistance Programme to the Department of Chemistry, Dibrugarh University.

REFERENCES

[1]   Bakherad, M.; Rezaeimanesh, F.; Nasr-Isfahani, H. Copper-Catalyzed Click Synthesis of Novel1,2,3-Triazole-Linked Pyrimidines. *ChemistrySelect,* **2018**, *3*, 2594-2598.
[http://dx.doi.org/10.1002/slct.201703088]

[2]   Maddila, S.; Pagadala, R.; Jonnalagadda, S. 1,2,4-Triazoles: A Review of Synthetic Approaches and the Biological Activity. *Lett. Org. Chem.,* **2013**, *10*, 693-714.
        [http://dx.doi.org/10.2174/1570178610101311126115448]

[3]   Hong, V.; Presolski, S.I.; Ma, C.; Finn, M.G. Analysis and Optimization of Copper-Catalyzed Azide–Alkyne Cycloaddition for Bioconjugation Angewandte Chemie (International ed. in English) , **2009**; 48, pp. 9879-9883.

[4]   Kolb, H.C.; Sharpless, K.B. The growing impact of click chemistry on drug discovery. *Drug Discov. Today,* **2003**, *8*(24), 1128-1137.
        [http://dx.doi.org/10.1016/S1359-6446(03)02933-7] [PMID: 14678739]

[5]   Wolfgang, H.B.; Christian, K. Azide/Alkyne-"Click" Reactions: Applications in Material Science and Organic Synthesis. *Curr. Org. Chem.,* **2006**, *10*, 1791-1815.
        [http://dx.doi.org/10.2174/138527206778249838]

[6]   Ma, L.; Leng, T.; Wang, K.; Wang, C.; Shen, Y.; Zhu, W. A coumarin-based fluorescent and colorimetric chemosensor for rapid detection of fluoride ion. *Tetrahedron,* **2017**, *73*, 1306-1310.
        [http://dx.doi.org/10.1016/j.tet.2017.01.034]

[7]   Agalave, S.G.; Maujan, S.R.; Pore, V.S. Click chemistry: 1,2,3-triazoles as pharmacophores. *Chem. Asian J.,* **2011**, *6*(10), 2696-2718.
        [http://dx.doi.org/10.1002/asia.201100432] [PMID: 21954075]

[8]   Huisgen, R. 1,3-Dipolar Cycloaddition Chemistry. New York: Wiley; , **1984**; pp. 1-176.

[9]   Rostovtsev, V.V.; Green, L.G.; Fokin, V.V.; Sharpless, K.B. A stepwise huisgen cycloaddition process: copper(I)-catalyzed regioselective "ligation" of azides and terminal alkynes. *Angew. Chem. Int. Ed.,* **2002**, *41*(14), 2596-2599.
        [http://dx.doi.org/10.1002/1521-3773(20020715)41:14<2596::AID-ANIE2596>3.0.CO;2-4] [PMID: 12203546]

[10]   Tornøe, C.W.; Christensen, C.; Meldal, M. Peptidotriazoles on solid phase: [1,2,3]-triazoles by regiospecific copper(i)-catalyzed 1,3-dipolar cycloadditions of terminal alkynes to azides. *J. Org. Chem.,* **2002**, *67*(9), 3057-3064.
        [http://dx.doi.org/10.1021/jo011148j] [PMID: 11975567]

[11]   Ali, A.A.; Sharma, R.; Saikia, P.J.; Sarma, D. CTAB promoted CuI catalyzed green and economical synthesis of 1,4-disubstituted-1,2,3-triazoles. *Synth. Commun.,* **2018**, *48*, 1206-1212.
        [http://dx.doi.org/10.1080/00397911.2018.1439176]

[12]   Wang, Q.; Chan, T.R.; Hilgraf, R.; Fokin, V.V.; Sharpless, K.B.; Finn, M.G. Bioconjugation by copper(I)-catalyzed azide-alkyne [3 + 2] cycloaddition. *J. Am. Chem. Soc.,* **2003**, *125*(11), 3192-3193. [3 + 2].
        [http://dx.doi.org/10.1021/ja021381e] [PMID: 12630856]

[13]   Garg, A.; Ali, A.A.; Damarla, K.; Kumar, A.; Sarma, D. Aqueous bile salt accelerated cascade synthesis of 1,2,3-triazoles from arylboronic acids. *Tetrahedron Lett.,* **2018**, *59*, 4031-4035.
        [http://dx.doi.org/10.1016/j.tetlet.2018.09.064]

[14]   Liang, L.; Astruc, D. The copper(I)-catalyzed alkyne-azide cycloaddition (CuAAC) "click" reaction and its applications. An overview. *Coord. Chem. Rev.,* **2011**, *255*, 2933-2945.
        [http://dx.doi.org/10.1016/j.ccr.2011.06.028]

[15]   Barman, M.K.; Sinha, A.K.; Nembenna, S. An efficient and recyclable thiourea-supported copper(i) chloride catalyst for azide–alkyne cycloaddition reactions. *Green Chem.,* **2016**, *18*, 2534-2541.
        [http://dx.doi.org/10.1039/C5GC02545A]

[16]   Dabiri, M.; Kasmaei, M.; Salari, P.; Movahed, S.K. Copper nanoparticle decorated three dimensional graphene with high catalytic activity for Huisgen 1,3-dipolar cycloaddition. *RSC Advances,* **2016**, *6*, 57019-57023.
        [http://dx.doi.org/10.1039/C5RA25317A]

[17]   Nemati, F.; Heravi, M.M.; Elhampour, A. Magnetic nano-Fe$_3$O$_4$@TiO$_2$/Cu$_2$O core–shell composite: an efficient novel catalyst for the regioselective synthesis of 1,2,3-triazoles using a click reaction. *RSC Advances,* **2015**, *5*, 45775-45784.
[http://dx.doi.org/10.1039/C5RA06810J]

[18]   Sabaqian, S.; Nemati, F.; Heravi, M.M.; Nahzomi, H.T. Copper(I) iodide supported on modified cellulose-based nano-magnetite composite as a biodegradable catalyst for the synthesis of 1,2,3-triazoles. *Appl. Organomet. Chem.,* **2017**, *31*, e3660.
[http://dx.doi.org/10.1002/aoc.3660]

[19]   Baig, R.B.N.; Varma, R.S. Copper on chitosan: a recyclable heterogeneous catalyst for azide–alkyne cycloaddition reactions in water. *Green Chem.,* **2013**, *15*, 1839.
[http://dx.doi.org/10.1039/c3gc40401c]

[20]   Maleki, A.; Taheri-Ledari, R.; Soroushnejad, M. Surface functionalization of magnetic nanoparticles *via* palladium-catalyzed Diels-Alder approach. *ChemistrySelect,* **2018**, *3*, 13057-13062.
[http://dx.doi.org/10.1002/slct.201803001]

[21]   Maleki, A.; Zand, P.; Mohseni, Z. Fe$_3$O$_4$@PEG-SO$_3$H rod-like morphology along with the spherical nanoparticles: novel green nanocomposite design, preparation, characterization and catalytic application. *RSC Advances,* **2016**, *6*, 110928-110934.
[http://dx.doi.org/10.1039/C6RA24029A]

[22]   Maleki, A.; Hajizadeh, Z.; Salehi, P. Mesoporous halloysite nanotubes modified by CuFe$_2$O$_4$ spinel ferrite nanoparticles and study of its application as a novel and efficient heterogeneous catalyst in the synthesis of pyrazolopyridine derivatives. *Sci. Rep.,* **2019**, *9*(1), 5552.
[http://dx.doi.org/10.1038/s41598-019-42126-9] [PMID: 30944394]

[23]   Chetia, M.; Gehlot, P.S.; Kumar, A.; Sarma, D. A recyclable/reusable hydrotalcite supported copper nanocatalyst for 1,4-disubstituted-1,2,3-triazole synthesis *via* click chemistry approach. *Tetrahedron Lett.,* **2018**, *59*, 397-401.
[http://dx.doi.org/10.1016/j.tetlet.2017.12.051]

[24]   Geim, A.K.; Novoselov, K.S. The rise of graphene. *Nat. Mater.,* **2007**, *6*(3), 183-191.
[http://dx.doi.org/10.1038/nmat1849] [PMID: 17330084]

[25]   aStoller, M.D.; Park, S.; Zhu, Y.; An, J.; Ruoff, R.S. Graphene-based ultracapacitors. *Nano Lett.,* **2008**, *8*(10), 3498-3502.
[http://dx.doi.org/10.1021/nl802558y] [PMID: 18788793] bAvouris, P.; Chen, Z.; Perebeinos, V. Carbon-based electronics. *Nat. Nanotechnol.,* **2007**, *2*(10), 605-615.
[http://dx.doi.org/10.1038/nnano.2007.300] [PMID: 18654384] cGuo, P.; Song, H.; Chen, X. *Electrochem. Commun.,* **2009**, *11*, 1320-1324.
[http://dx.doi.org/10.1016/j.elecom.2009.04.036] dWang, X.; Zhi, L.; Tsao, N.; Tomovi, Z.; Li, J.; Mullen, K. *Angew. Chem. Int. Ed.,* **2008**, *47*, 2990-2992.
[http://dx.doi.org/10.1002/anie.200704909] eYoo, E. T.; Akita, T.; Kohyama, M.; Nakamura J.; Honma, I. Enhanced Electrocatalytic Activity of Pt Subnanoclusters on Graphene Nanosheet Surface. *Nano Lett.,* **2009**, *9*, 2255-2259.
[http://dx.doi.org/10.1021/nl900397t] [PMID: 19405511] fTang, L.; Wang, Y.; Li, Y.; Feng, H.; Lu, J.; Li, J. Preparation, Structure, and Electrochemical Properties of Reduced Graphene Sheet Films Adv. *Funct. Mater.,* **2009**, *19*, 2782-2789.
[http://dx.doi.org/10.1002/adfm.200900377]

[26]   Pasricha, R.; Gupta, S.; Srivastava, A.K. A facile and novel synthesis of Ag-graphene-based nanocomposites. *Small,* **2009**, *5*(20), 2253-2259.
[http://dx.doi.org/10.1002/smll.200900726] [PMID: 19582730]

[27]   Bai, S.; Shen, X.; Zhu, G.; Xu, Z.; Liu, Y. Reversible phase transfer of graphene oxide and its use in the synthesis of graphene-based hybrid materials. *Carbon,* **2011**, *49*, 4563-4570.
[http://dx.doi.org/10.1016/j.carbon.2011.06.072]

[28]   Kong, B.S.; Geng, J.; Jung, H.T. Layer-by-layer assembly of graphene and gold nanoparticles by vacuum filtration and spontaneous reduction of gold ions. *Chem. Commun. (Camb.),* **2009**, (16), 2174-2176.
[http://dx.doi.org/10.1039/b821920f] [PMID: 19360184]

[29]   Scheuermann, G.M.; Rumi, L.; Steurer, P.; Bannwarth, W.; Mülhaupt, R. Palladium nanoparticles on graphite oxide and its functionalized graphene derivatives as highly active catalysts for the Suzuki-Miyaura coupling reaction. *J. Am. Chem. Soc.,* **2009**, *131*(23), 8262-8270.
[http://dx.doi.org/10.1021/ja901105a] [PMID: 19469566]

[30]   Warner, J.H.; Rümmeli, M.H.; Bachmatiuk, A.; Wilson, M.; Büchner, B. Examining co-based nanocrystals on graphene using low-voltage aberration-corrected transmission electron microscopy. *ACS Nano,* **2010**, *4*(1), 470-476.
[http://dx.doi.org/10.1021/nn901371k] [PMID: 20020749]

[31]   Bin, X.; Chen, J.; Cao, H.; Chen, L.; Yuan, J. Preparation of graphene encapsulated copper nanoparticles from CUCl2-GIC. *J. Phys. Chem. Solids,* **2009**, *70*, 1-7.
[http://dx.doi.org/10.1016/j.jpcs.2007.10.015]

[32]   Wang, D.; Choi, D.; Li, J.; Yang, Z.; Nie, Z.; Kou, R.; Hu, D.; Wang, C.; Saraf, L.V.; Zhang, J.; Aksay, I.A.; Liu, J. Self-assembled TiO2-graphene hybrid nanostructures for enhanced Li-ion insertion. *ACS Nano,* **2009**, *3*(4), 907-914.
[http://dx.doi.org/10.1021/nn900150y] [PMID: 19323486]

[33]   Chen, S.; Zhu, J.; Wu, X.; Han, Q.; Wang, X. Graphene oxide--MnO$_2$ nanocomposites for supercapacitors. *ACS Nano,* **2010**, *4*(5), 2822-2830.
[http://dx.doi.org/10.1021/nn901311t] [PMID: 20384318]

[34]   Fu, W.; Zhang, Z.; Zhuang, P.; Shen, J.; Ye, M.J. Colloid Interface. *Forensic Sci.,* **2017**, *497*, 83.

[35]   Lonkar, S.P.; Ahmed, A.A. *J. Thermodyn. Catal.,* **2014**, *5*, 132.

[36]   Jain, A.; Jain, Y.; Gupta, R.; Agarwal, M. Trifluoromethyl group containing C 3 symmetric coumarin-triazole based fluorometric tripodal receptors for selective fluoride ion recognition: A theoretical and experimental approach. *J. Fluor. Chem.,* **2018**, *212*, 153-160.
[http://dx.doi.org/10.1016/j.jfluchem.2018.06.005]

[37]   Shaygan Nia, A.; Rana, S.; Döhler, D.; Noirfalise, X.; Belfiore, A.; Binder, W.H. Click chemistry promoted by graphene supported copper nanomaterials. *Chem. Commun. (Camb.),* **2014**, *50*(97), 15374-15377.
[http://dx.doi.org/10.1039/C4CC07774A] [PMID: 25350638]

[38]   Reddy, V.H.; Reddy, Y.V.R.; Sridhar, B.; Reddy, B.V.S. Green Catalytic Process for Click Synthesis Promoted by Copper Oxide Nanocomposite Supported on Graphene Oxide. *Adv. Synth. Catal.,* **2016**, *358*, 1088-1092.
[http://dx.doi.org/10.1002/adsc.201501072]

[39]   Putta, C.B.; Sharavath, V.; Sarkar, S.; Ghosh, S. Palladium nanoparticles on β-cyclodextrin functionalised graphene nanosheets: a supramolecular based heterogeneous catalyst for C–C coupling reactions under green reaction conditions. *RSC Advances,* **2015**, *5*, 6652-6660.
[http://dx.doi.org/10.1039/C4RA14323J]

[40]   Jain, Y.; Kumari, M.; Singh, R.P.; Kumar, D.; Gupta, R. Sonochemical Decoration of Graphene Oxide with Magnetic Fe$_3$O$_4$@ CuO Nanocomposite for Efficient Click Synthesis of Coumarin-Sugar Based Bioconjugates and Their Cytotoxic Activity. *Catal. Lett.,* **2020**, *150*, 1142-1154.
[http://dx.doi.org/10.1007/s10562-019-02982-6]

[41]   Jain, Y.; Kumari, M.; Laddha, H. Gupta, Ragini. Ultrasound Promoted Fabrication of CuO-Graphene Oxide Nanocomposite for Facile Synthesis of Fluorescent Coumarin Based 1,4-disubstituted 1,2,3-triazoles in Aqueous Media. *ChemistrySelect,* **2019**, *4*, 7015-7026.
[http://dx.doi.org/10.1002/slct.201901355]

[42]   Salam, N.; Sinha, A.; Mondal, P.; Roy, A.S.; Jana, N.R.; Islam, S.M. Efficient and reusable graphene-γ-Fe$_2$O$_3$ magnetic nanocomposite for selective oxidation and one-pot synthesis of 1,2,3-triazole using green solvent. *RSC Advances,* **2013**, *3*, 18087-18098.
[http://dx.doi.org/10.1039/c3ra43184c]

[43]   Ji, Z.; Shen, X.; Zhu, G.; Zhou, H.; Yuan, A. Reduced graphene oxide/nickel nanocomposites: facile synthesis, magnetic and catalytic properties. *J. Mater. Chem.,* **2012**, *22*, 3471-3477.
[http://dx.doi.org/10.1039/c2jm14680k]

[44]   Harmata, M. *Silver in Organic Chemistry*; John Wiley & Sons: Hoboken, **2010**.
[http://dx.doi.org/10.1002/9780470597521]

[45]   aPalaniappan, S.; Rajender, B. A Novel Polyaniline-Silver Nitrate-*p*-Toluenesulfonic Acid Salt as Recyclable Catalyst in the Stereoselective Synthesis of β-Amino Ketones: "One-Pot" Synthesis in Water Medium. *Adv. Synth. Catal.,* **2010**, *352*, 2507.
[http://dx.doi.org/10.1002/adsc.201000346] bBaby, T.T.; Ramaprabhu, S. Synthesis and nanofluid application of silver nanoparticles decorated graphene. *J. Mater. Chem.,* **2011**, *21*, 9702.
[http://dx.doi.org/10.1039/c0jm04106h] cShen, J.; Shi, M.; Yan, B.; Ma, H.; Li, N.; Ye, M. One-pot hydrothermal synthesis of Ag-reduced graphene oxide composite with ionic liquid. *J. Mater. Chem.,* **2011**, *21*, 7795.
[http://dx.doi.org/10.1039/c1jm10671f] dTang, X.Z.; Cao, Z.; Zhang, H.B.; Liu, J.; Yu, Z.Z. Growth of silver nanocrystals on graphene by simultaneous reduction of graphene oxide and silver ions with a rapid and efficient one-step approach. *Chem. Commun. (Camb.),* **2011**, *47*(11), 3084-3086.
[http://dx.doi.org/10.1039/c0cc05613h] [PMID: 21298137]

[46]   Salam, N.; Sinha, A.; Roy, A.S.; Mondal, P.; Jana, N.R.; Islam, S.M. Synthesis of silver–graphene nanocomposite and its catalytic application for the one-pot threecomponent coupling reaction and one-pot synthesis of 1,4-disubstituted 1,2,3-triazoles in water. *RSC Advances,* **2014**, *4*, 10001-10012.
[http://dx.doi.org/10.1039/c3ra47466f]

[47]   Xu, C.; Wang, X.; Yang, L.; Wu, Y. Fabrication of ZnS-Bi-TiO$_2$ Composites and Investigation of Their Sunlight Photocatalytic Performance. *J. Solid State Chem.,* **2009**, *182*, 2486-9240.
[http://dx.doi.org/10.1016/j.jssc.2009.07.001]

[48]   Roy, I.; Bhattacharyya, A.; Sarkar, G.; Saha, N.R.; Rana, R.; Ghosh, P.P.; Palit, M. Das. A. R.; Chattopadhyay, D. *In situ* synthesis of a reduced graphene oxide/cuprous oxide nanocomposite: a reusable catalyst. *RSC Advances,* **2014**, *4*, 52044-52052.
[http://dx.doi.org/10.1039/C4RA08127G]

[49]   Pourjavadia, A.; Safaie, N.; Hosseini, S.H.; Bennett, C. Graphene oxide/poly(vinyl imidazole) nanocomposite: an effective support for preparation of highly loaded heterogeneous copper catalyst. *Appl. Organomet. Chem.,* **2015**, *29*, 601-607.
[http://dx.doi.org/10.1002/aoc.3336]

[50]   Diyarbakir, S.; Can, H.; Metin, Ö. Reduced graphene oxide-supported CuPd alloy nanoparticles as efficient catalysts for the Sonogashira cross-coupling reactions. *ACS Appl. Mater. Interfaces,* **2015**, *7*(5), 3199-3206.
[http://dx.doi.org/10.1021/am507764u] [PMID: 25594280]

[51]   Chen, M.X.; Zhang, Z.; Li, L.Z.; Liu, Y.; Wang, W.; Gao, J.P. Fast synthesis of Ag-Pd@reduced graphene oxide bimetallic nanoparticles and their applications as carbon–carbon coupling catalysts. *RSC Advances,* **2014**, *4*, 30914.
[http://dx.doi.org/10.1039/C4RA05186F]

[52]   Fu, W.; Zhang, Z.; Zhuang, P.; Shen, J.; Ye, M.J. One-pot hydrothermal synthesis of magnetically recoverable palladium/reduced graphene oxide nanocomposites and its catalytic applications in cross-coupling reactions. *Colloid InterfaceSci.,* **2017**, *497*, 83.
[http://dx.doi.org/10.1016/j.jcis.2017.02.063]

[53]   Guo, X.; Hao, C.; Jin, G.; Zhu, H.Y.; Guo, X.Y. Copper nanoparticles on graphene support: an

efficient photocatalyst for coupling of nitroaromatics in visible light. *Angew. Chem. Int. Ed. Engl.,* **2014**, *53*(7), 1973-1977.
[http://dx.doi.org/10.1002/anie.201309482] [PMID: 24505013]

[54]   Li, Z.; Zhao, H.; Han, H.; Song, J.; Liu, Y.; Guo, W.; Sun, Z. W. A one-pot method for synthesis of reduced graphene oxide- supported Cu–Cu2O and catalytic application in tandem reaction of halides and sodium azide with terminal alkynes. *Appl Organometal Chem ,* **2018**, e4301.

[55]   Ashori, A.; Cordeiro, N.; Faria, M.; Hamzeh, Y. Effect of chitosan and cationic starch on the surface chemistry properties of bagasse paper. *Int. J. Biol. Macromol.,* **2013**, *58*, 343-348.
[http://dx.doi.org/10.1016/j.ijbiomac.2013.04.056] [PMID: 23624167]

[56]   Reddy, D.H.K.; Lee, S.M. *Application of magnetic chitosan composites for the removal of toxic metal and dyes from aqueous solutions, Adv*; Colloid Interf. Sci, **2013**, pp. 20168-20193.

[57]   Mahdavinasab, M.; Hamzehloueian, M.; Sarrafi, Y. Preparation and application of magnetic chitosan/graphene oxide composite supported copper as a recyclable heterogeneous nanocatalyst in the synthesis of triazoles. *Int. J. Biol. Macromol.,* **2019**, *138*, 764-772.
[http://dx.doi.org/10.1016/j.ijbiomac.2019.07.013] [PMID: 31284011]

[58]   Aghayan, M.M.; Saeedi, M. Boukherroub. R. Cu2O/reduced graphene oxide/TiO2 nanomaterial: An effective photocatalyst for azide-alkyne cycloaddition with benzyl halides or epoxide derivatives under visible light irradiation. *Appl. Organomet. Chem.,* **2020**, 5928.

[59]   Rafiee, F.; Khavari, P. Preparation of aryl azides of aryl boronic acids and one-pot synthesis of 1,4-diaryl-1,2,3-triazoles by a magnetic cysteine functionalized GO–CuI/II nanocomposite. *Appl. Organomet. Chem.,* **2020**, 5789.
[http://dx.doi.org/10.1002/aoc.5789]

CHAPTER 6

Carbon Dioxide Conversion to Value-Added Chemicals using Graphene and its Composite Materials

Sandeep Kumar[1,*], Nayuesh Sharma[1], Ajit Sharma[2] and Deepak Kumar[2,*]

[1] Department of Chemistry, Indian Institute of Technology Ropar, Rupnagar 140001, India

[2] Department of Chemical Engineering and Physical Sciences, Lovely Professional University, Phagwara 144411, India

Abstract: Rapidly growing industrialization leads to an increase in global temperature as a result of the emission of greenhouse gases, such as carbon dioxide (CO_2), in the environment. CO_2 is an abundant, nontoxic, and renewable C1 feedstock source of carbon for synthesizing fine chemicals. CO_2 can be transformed into many gaseous or liquid fuels and fine chemicals, such as carbon monoxide, methane, formic acid, formaldehyde, methanol, cyclic carbonates, and other hydrocarbon fuels. The inherent thermodynamics stability and kinetic inertness of CO_2 pose a limit for the synthesis of fine chemicals. Various materials have been explored by many researchers around the globe to generate efficient chemicals and products from CO_2. Graphene and its composite materials have a large surface area, and their surface is rich in various reactive oxygen functional groups (-C-OH, -C=O, -COOH, *etc.*), which can be highly active sites for various organic catalytic reactions, including oxidation, reduction, ring-opening, and coupling reactions. In this chapter, the structure, strategies, and mechanism of utilizing CO_2 into value-added chemicals using graphene and its composite materials will be presented.

Keywords: Catalytic reaction, Carbon dioxide reduction, Carbon monoxide, Cyclic carbonates, Formic acid, Formaldehyde, Graphene, Graphite, Hydrocarbon fuels, Methane, Methanol, Value-added chemicals.

INTRODUCTION

An increase in energy demand is observed due to the rapid growth of the world population and changes in lifestyle. Presently, the use of petroleum-based fossil fuels (*viz.* natural gas, coal, and oil) and growing industrialization lead to an incr-

---
[*] **Corresponding author Sandeep Kumar and Deepak Kumar:** Department of Chemistry, Indian Institute of Technology Ropar, Rupnagar 140001, India and Department of Chemical Engineering and Physical Sciences, Lovely Professional University, Phagwara 144411, India;
E-mails: sandeep.kumar@iitrpr.ac.in, deepak.sharma99967@gmail.com

Manorama Singh, Vijai K. Rai and Ankita Rai (Eds)

ease in overall temperature as a result of the emission of greenhouse gases, such as carbon dioxide (Co_2) content to the environment, which results in global warming [1].

The Intergovernmental Panel on Climate Change (IPCC) estimated that the atmospheric CO_2 concentration would rise up to 950 ppm by the end of the year 2100 [2]. To eliminate these consequences and minimize the atmospheric CO_2 emission, the capture and conversion of CO_2 into value-added chemicals are highly demanded. CO_2 is an abundant, nontoxic, and renewable source of carbon (C1 feedstock) for the development of efficient and value-added chemicals [3]. Many useful chemicals, *viz.* carbon monoxide, methane, formic acid, formaldehyde, methanol, cyclic carbonates, and hydrocarbon fuels, can be synthesized by CO_2 conversion [4, 5]. Therefore, various strategies have been developed to convert CO_2 into value-added chemicals. The homogeneous catalytic systems, such as metal complexes, salen complexes, porphyrin complexes, ionic liquids, and organic compounds, have fewer advantages due to intricate difficulty in the catalyst-recycling and low surface area, limiting their practical and large-scale industrial applications [6].

In this context, graphene and its modified composites as heterogeneous catalysts can effectively overcome the catalyst recycling problem in homogeneous catalytic systems. Graphene is a flat monolayer of tightly packed sp^2 hybridized carbon atoms into a two-dimensional (2D) honeycomb lattice [7, 8]. It is a basic building block of all graphitic materials of different dimensional structures. Graphene can be molded into 0D fullerenes, spin into 1D nanotubes, and changed into 3D graphite by stacking. Furthermore, graphene and its composite materials have a large surface area, high carrier mobility, high thermal conductivity, and their surface is rich in various reactive oxygen functional groups (-C-OH, -C=O, -COOH, *etc.*), which can be highly active sites for various organic catalytic reactions including oxidation, reduction, ring-opening, and coupling reactions [6, 9 - 11]. Various methods have been studied and reported for the synthesis of graphene and graphene-based analogous, such as chemical exfoliation of graphite, chemical vapor deposition, and solvothermal method [6]. Many reports have shown that pristine graphene has neutral carbon atoms and is not useful for CO_2 reduction. Various theoretical and experimental reports have shown that heteroatom (such as boron, nitrogen, oxygen, and phosphorous) doped graphene incorporated in graphene changes its chemical reactivity and electronic properties [10, 11]. These hybrid materials enhance its catalytic activity against CO_2 reduction. The rational design and morphology of graphene-based materials play an important role in the CO_2 reduction reaction. The bare graphene and its composite materials have different properties.

Graphene has excellent properties such as high electron transport at room temperature, good mechanical properties, magnificent thermal conductivity, *etc.* However, the band gap of graphene can be modified into semiconducting material to be used as a catalyst and transistors. Graphene can be further modified by its shape, size, and chemical structure based on the applications [10, 11]. In this chapter, the structure, strategies, and mechanism of utilizing CO_2 into value-added chemicals using graphene and its composite materials will be presented.

GRAPHENE-BASED COMPOSITE MATERIALS FOR CO_2 CONVERSION TO CYCLIC CARBONATES

Cyclic carbonates are very useful compounds at the industrial level as they are used as reagents, solvents, diluents, and monomers for polymers. Among the various reactions of CO_2, the cycloaddition reaction of CO_2 with epoxides to generate cyclic carbonates has attracted a special interest due to the high yield and selectivity of cyclic carbonates without producing any by-products [12 - 19]. However, the high thermodynamic stability and kinetic inertness of CO_2 put a great challenge for its utilization under mild conditions. An efficient catalyst system for cycloaddition of carbon dioxide with epoxides to generate cyclic carbonates should possess a high density of CO_2-philic basic sites and Lewis/Bronsted acidic catalytic sites along with a high affinity for carbon dioxide [20 - 23]. In this context, various catalyst systems have been developed for cycloaddition of CO_2 with epoxides to generate cyclic carbonates. Among them, graphene-based composite materials have gained a special interest owing to their 2D structure, which confers a high surface area and introduces a high density of acidic and basic CO_2-philic functionalities for applications of selective capture and conversion of carbon dioxide. However, most of the catalyst systems known for the cycloaddition of carbon dioxide with epoxides require an additional co-catalyst for the high yield generation of cyclic carbonates. Interestingly, the use of graphene-based composite materials is beneficial in introducing both basic and acidic sites in a single system for catalysis, leading to selective capture and conversion of CO_2. In addition, the presence of several oxygen-containing groups (–OH, –COOH, *etc.*) leads to a synergistic effect activating the CO_2 molecules through hydrogen bond interactions with epoxide substrates. The graphene-based composite materials can activate the epoxide ring by coordinating the Lewis acidic sites (M) followed by ring-opening of the epoxide from less hindered carbon atom by nucleophilic attack of the Br^- anion from the co-catalyst such as TBAB to generate a bromoalkoxide (Fig. **1**). Subsequently, the polarization of CO_2 takes place at the functional groups to generate polarized CO_2 molecules, followed by insertion into the alkoxide bond, which results in the formation of carbonate species which upon intramolecular ring-closure reaction yields cyclic carbonate. Subsequent elimination of the cyclic carbonate from the catalyst leads to regeneration of the active catalyst and makes the cycle continues, as shown in Fig. (**1**)

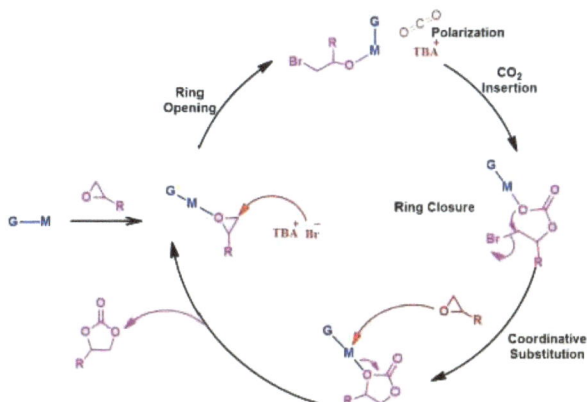

Fig. (1). A plausible reaction mechanism for the cycloaddition of CO_2 with epoxides catalysed by graphene-based composite materials.

In this context, Qu and co-workers demonstrated the use of commercially available graphene oxide (GO) as a carbocatalyst in the presence of DMF for the generation of cyclic carbonates [24]. They showed that the number of oxygenated groups is proportional to the catalytic activity of GO. Furthermore, GO can be modified using oxygenated groups as reactive species of proper silanes. Towards this direction, Bhanage and co-workers synthesized 3-aminopropyl-grafted GO by condensation reaction between 3-aminopropyltrimethoxysilane and GO. This hybrid was used as a catalyst with additional co-catalysts such as tetrabutylammonium iodide (TBAI) to convert CO_2 into cyclic carbonates [25]. The synergistic effect between the oxygenated GO and amine moieties leads to hydrogen-bonding donor sites that effectively activate CO_2 molecules and epoxides. This heterogeneous catalyst is so efficient that only 0.1 MPa atmospheric pressure of CO_2 is required for its conversion into the corresponding cyclic carbonates within 27 h at 100 °C.

Remarkably, increasing CO_2 pressure from 0.1 MPa to 1 MPa reduces the reaction temperature and time to 70 °C and 12 h, respectively. This result shows that the pressure of CO_2 plays an important role in determining the kinetics of this reaction. In addition, this heterogeneous catalyst is recyclable for up to seven cycles, demonstrating the reusability of this catalyst.

Furthermore, Yin and co-workers reported one-pot synthesis for the immobilization of a series of materials such as silanol group, multi-cationic quaternary ammonium salt, and tertiary amine over graphene oxide (GO) materials [26]. This was the first time when a multifunctionalized GO (MF-GO) enriched with multi-cationic quaternary ammonium salt was synthesized and utilized to convert CO_2 into the corresponding cyclic carbonates under mild conditions with high efficiency without the need of a solvent and a co-catalyst. They proposed the mechanism of this conversion using propylene oxide (PO) as a model substrate; first, PO is coordinated with the Si-OH on GO-H-Me through hydrogen bonding interactions along with the adsorption and activation of CO_2 by ternary amine groups leading to the formation of carbamate species inside the cage of H-atom. Then the nucleophilic I anions attack the less sterically hindered carbon atom of the activated PO followed by ring-opening to generate the oxygen anion, which is further stabilized through hydrogen-bonding interactions with the Si-OH group. Furthermore, the activated CO_2 molecules undergo an insertion reaction with obtained oxygen anion, leading to the formation of an alkyl carbonate anion, which is also stabilized by Si-OH. Finally, corresponding cyclic carbonates are generated through an intramolecular cyclic step along with the release of a regenerated catalyst. Here, a combination of the "multi-cationic approach" and "multi-synergetic strategy" was successfully presented for the design of a suitable catalyst system for the conversion of CO_2 for cyclic carbonates under mild conditions. Interestingly, the MF-GO catalyst can be recyclable for up to five cycles without significant loss of catalytic activity.

Xu *et al.* reported the preparation of a series of imidazolium-based ionic liquids doped onto the surface of GO [27]. The hybrid catalyst efficiently converted CO_2 at 2 MPa pressure into cyclic carbonates and was recyclable without losing catalytic activity. In a hybrid catalyst, the iodide counter anions as nucleophilic co-catalyst first open the epoxide ring, followed by the insertion of CO_2 to generate the desired product. It is worth mentioning that higher conversions of propylene oxide into the corresponding carbonate were obtained using the residual hydroxyl groups of GO, which acted as hydrogen-bond donors since their silylation resulted in a marked drop in the catalytic activity [28]. This synergistic effect in accelerating the ring-opening of epoxides ring by hydroxyl groups of GO was further confirmed with the use of catalyst, in which a hydroxyl functionalized ionic liquid was immobilized onto GO, showing the beneficial effect of hydroxyl groups in the imidazolium tag. Interestingly, the catalyst could be recycled up to seven times without any significant loss of catalytic activity.

GRAPHENE-BASED COMPOSITE MATERIALS FOR PHOTOCATALYTIC CONVERSION OF CO_2

Semiconductor photocatalysts are materials for bio-mimicking natural photosynthesis for direct conversion of solar energy to chemical energy [29]. In particular, in 1979, the Honda group first time reported the photocatalytic reduction of CO_2 for the generation of methane, methanol, formaldehyde, and formic acid, which are high energy density fuels (*e.g.*, methane,methanol) using aqueous suspensions of semiconductor photocatalysts such as TiO_2, ZnO, CdS, GaP, and SiC (Fig. **2**) [30, 31]. Over the last few decades, continuous efforts have been made for the construction of new photocatalytic materials, especially for the photocatalytic reduction of CO_2 to produce solar fuels and chemicals. Although significant progress has been made in the construction of new photocatalytic materials, these photocatalysts still suffer from some severe limitations such as inferior utilization of solar light, low quantum efficiency, instability of some photocatalysts (*e.g.*, metal sulfides), and difficulty in controlling the product selectivity of some reactions (*e.g.*, photocatalytic reduction of CO_2). Various strategies have been adopted to overcome these challenges, such as doping of metal ions and complexes, polymers, metal-organic frameworks, nanomaterial modification, *etc.* However, in recent years, the fabrication of graphene-based composite materials has made it a highly promising material for improving solar energy conversion efficiency due to its large specific surface area, good optical properties, high work function, and exceptional electronic conductivity of graphene [30, 32]. In this regard, Chai and co-workers reported the generation of nitrogen-doped-TiO_2-001/graphene (N-TiO_2-001/GR) composites *via in situ* growth of well-faceted N-TiO_2 with exposed (001) facets onto GR sheets for photocatalytic reduction of CO_2 to CH_4 in the presence of water vapor (Fig. **3**) [33].

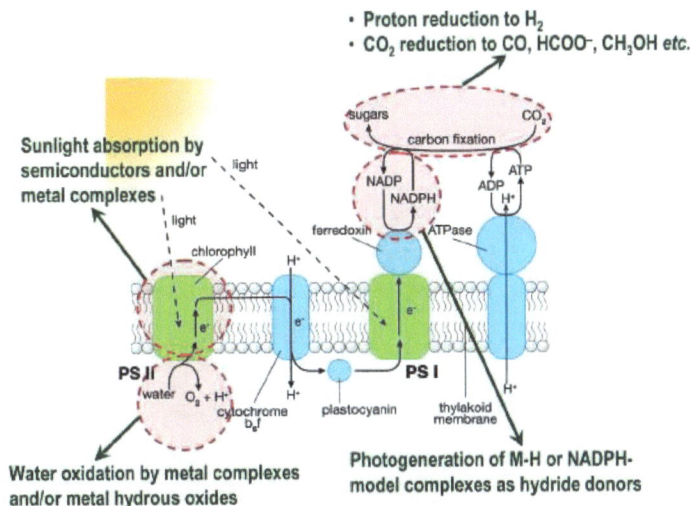

Fig. (2). Schematic diagram of a natural photosynthetic system, with four areas of artificial photocatalysis research, highlighted in red and described in green text. The artificial photocatalytic systems are designed to perform similar functions to the different components of natural photosystems, which typically focus on one or two of these topics at a time to reduce complexity. The goal of photocatalytic reduction of CO_2 is to mimic the natural photosynthetic system for the generation of renewable fuels and chemicals such as CO, formic acid, methanol, and methane from CO_2, H_2O, and sunlight [31].

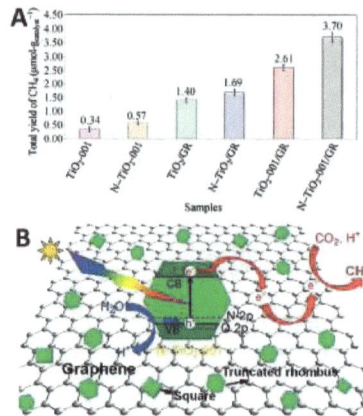

Fig. (3). The total yield of CH_4 over as-synthesized TiO_2-001 and N-TiO_2-001 and a series of TiO_2-based/GR composites (A); schematic illustration of the charge transfer and separation of electron-hole pairs for the reduction of CO_2 with H_2O to CH_4 using N-TiO_2-001/GR composites under visible light irradiation (B) [33].

Furthermore, Yong and co-workers demonstrated the fabrication of a series of sandwich-like graphene–(g-C_3N_4) (GCN) composites by a one-pot impregnation–thermal reduction strategy for the conversion of CO_2 to CH_4 using water vapor as a hole scavenger [34]. Furthermore, Baeg and co-workers synthesized photocatalytic graphene-based composite composed of chemically converted graphene coupled with multi-anthraquinone substituted porphyrin (CCGCMAQSP), an organometallic rhodium complex, nicotinamide adenine dinucleotide (NADH), and formate dehydrogenase. The designed composite shows photocatalytic reduction of CO_2 to formic acid (HCOOH) in the presence of triethanolamine (TEOA) as a sacrificial electron donor (Fig. **4**) [35].

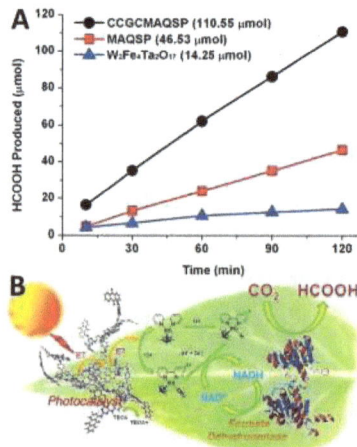

Fig. (4). Photocatalytic activities of CCGCMAQSP, MAQSP, and $W_2Fe_4Ta_2O_{17}$ in visible-light-driven artificial photosynthesis of formic acid from Co_2 (A); schematic illustration of artificial photosynthesis of formic acid from CO_2 over a graphene-based photocatalyst catalyzed under visible light irradiation (B) [35].

The conversion of CO_2 to methanol has significant importance because methanol is widely used as a solvent for the synthesis of various organic compounds. In this context, Jain *et al.* reported the visible light-assisted photocatalytic reduction of CO_2 to methanol in DMF/H_2O solution in the presence of triethylamine as a sacrificial electron donor using GO immobilized with metal-organic complexes (Ru-(phen-GO).

In this composite, they immobilized ruthenium trinuclear polyazine complexes GO, which was modified with phenanthroline. Remarkably, Ru-(phen-GO) composite displayed efficient visible light-assisted photoactivity compared to parent GO for photocatalytic reduction of CO_2 to methanol (Fig. **5**) [36]. The higher photocatalytic reduction of Ru-(phen-GO) composite over parent GO is ascribed to strongly absorb visible light property of ruthenium, which injects electrons into the conduction band of GO; therefore, subsequently, adsorbed CO_2 is reduced to methanol on the surface of GO. Meanwhile, the electron deficiency in the oxidized ruthenium is filled by triethylamine as a sacrificial electron donor, which allows the continuation of the process.

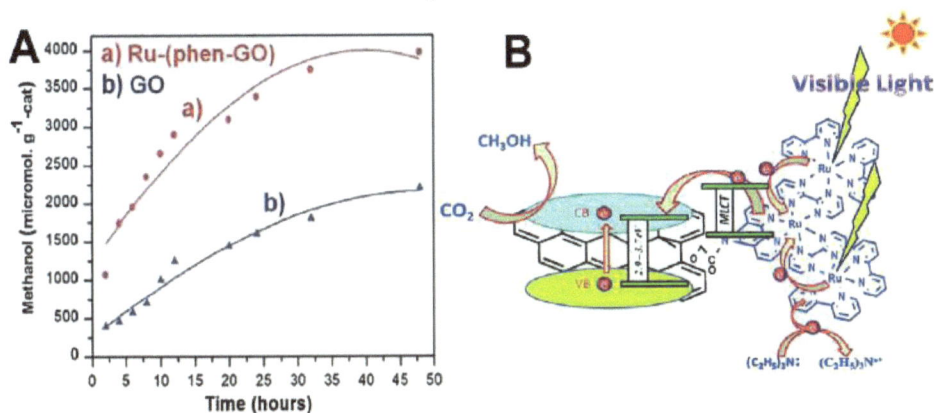

Fig. (5). Photocatalytic conversion of CO_2 to methanol over the Ru–(phen–GO) composites and GO (A) and the possible mechanism of photocatalytic Co_2 reduction over Ru–(phen–GO) composites (B) [36].

GRAPHENE-BASED COMPOSITE MATERIALS FOR CO_2 CONVERSION TO HYDROCARBON FUELS

CO_2 conversion into value-added chemicals by the photocatalytic process is an efficient and clean pathway for converting solar energy into sustainable and clean fuel and chemicals. Considering today's increasing energy demand and environmental pollution, conversion of CO_2 into hydrocarbon fuel is helpful to bring down the CO_2 from the atmosphere and also helps in fulfilling the energy requirement.

Wu and co-workers showed the selective conversion of CO_2 to C_2H_6 through the multi-electron process for artificial photosynthesis on graphene-modified chlorophyll Cu [37]. The effective photocatalyst chlorophyll-a/graphene composite (g-Chl-a) was produced from graphene and chlorophyll a (Chl-a), extracted from silkworm excrement. Photocatalyst g-Chl-a was used for CO_2 reduction reaction (CO_2RR) by the multi-electron process. Three types of C2 hydrocarbons, namely, C_2H_2, C_2H_4, and C_2H_6, were generated on chlorophyll-a/graphene composite. However, to increase the selectivity and stability, the substitution of central Mg^{2+} in Chl-a by Cu^{2+} cation was performed using copper acetate to synthesize Chl-Cu ($C_{55}H_{72}O_5N_4Cu$). Composite g-Chl-Cu showed improved selectivity and stability and produced only C_2H_6 with 68.23 µmol m^{-2} h^{-1} of production rate. The collaborative effect between graphene and Chl-Cu acted as an oxidation and reduction center, respectively, in the photocatalytic reduction of CO_2 into C_2H_6.

Recent work has shown that the holes on the valence band of Chl-Cu can oxidize H_2O to O_2, while electrons in the conduction band of Chl-Cu suggest the reduction of CO_2 to low valence products, indicating photocatalysis on Chl-Cu [37]. The excitons transferred from the chlorophyll antenna of the g-Chl-Cu complex to the CO_2 molecules. First, the electrons were excited from the valence band to the conduction band of Chl-Cu by light, then generated electrons were transferred to the conduction band of graphene, and after that, H_2O molecules which were adsorbed on the porphyrin group oxidized to O_2 and in the end, transferred electrons were reacted with the CO_2 molecules adsorbed on graphene, as shown in Fig. (**6**).

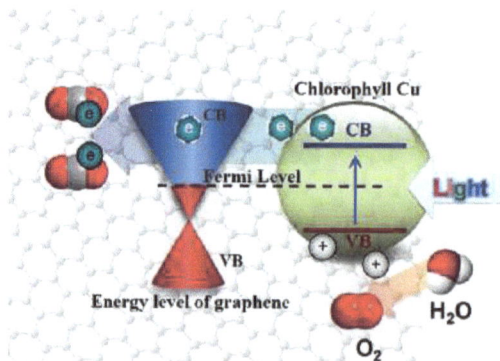

Fig. (6). The potential electron transfer process among Chl-Cu, graphene, H_2O, and CO_2 [37].

Here, photocatalysis was followed by two mechanisms: generation of electrons by light absorption and using the excited electrons by an electrocatalytic process. Wu *et al.* showed that the reduction of CO_2/CO dimer on the catalyst to produce ethylene and C_2H_6. The proton-assisted multiple-electron transfer is the most convenient and alternative route to reduce Co_2 by CO/Co_2 coupling route. In CO/CO_2 coupling pathway with g-Chl-a as a catalyst, a small quantity of C_2H_2 and C_2H_4 were generated, which was regular with this pathway because C_2H_2, C_2H_4, C_2H_6, and C_2H_5OH were tended to generate on metal or metal oxide catalyst [38, 39].

Here, in this photosynthesis, water behaves as an electron donor ($2H_2O \rightarrow 2H^+ + H_2O_2 + 2e^-$; $H_2O_2 \rightarrow 2H^+ + O_2 + 2e^-$; $H_2O_2 \rightarrow H_2O + O_2$; $2H_2O \rightarrow 4H^+ + O_2 + 4e^-$) and donated electrons. It is the main source of electrons for the CO_2 reduction reaction. H_2O oxidization is the rate-limiting step in this multi-electron reaction. For further reduction, it makes available multi-electrons and multi-protons. Multielectron supports C–C coupling. Amatore *et al.* reported the possible mechanism through the self-coupling of CO_2 in the presence of multi-electrons and multi-protons to form the oxalate in the first step [40]. The formation of oxalate is a crucial step in the conversion of CO_2 to C_2 products, as shown in Fig. (7). The second step is the formation of Co_2 from oxalate *via* a series of reactions, *viz.* dehydration of a vicinal diol, tautomerization of a ketone/enol, and dehydration of an enol. The final step is the formation of C_2H_6 by reducing C_2H_2 *via* hydrogenation reactions [41, 42]. All the intermediates are shown in Fig. (7) in their delocalized electronic structure. Intermediates are attached to the graphene sheet surface by a π–π non-covalent bond, which accepts electrons continuously until the formation of C_2H_6. Furthermore, the electrostatic interaction, charge dispersion effect, and polarization effect enhance the interaction between intermediates and graphene and increase the stability of the intermediates [43, 44]. The Chl-Cu showed high selectivity compared to the Chl-a because, after Cu^{2+} substitution, the potential changed in the photogenerated electrons [45].

Fig. (7). Proposed mechanistic steps in the reduction of CO_2 through a multi-electron process [37]

Wu and co-workers explored the product C_2H_6 by visible light irradiation. The photocatalyst was highly selective for CO_2 conversion [37]. The selectivity of photocatalyst for CO_2 reduction to one product was increased by controlling the electron transfer mechanism by tailoring the properties of Chl-Cu and graphene sheet in the photocatalyst. A possible and rational approach is suggested to explore the multi-electron pathway for a more efficient and selective photocatalytic CO_2 reduction reaction.

GRAPHENE-BASED COMPOSITE MATERIALS FOR ELECTROCHEMICAL CO_2 CONVERSION

Pristine graphene and graphene oxide in their natural form cannot be used for electrocatalytic CO_2 reduction as they have neutral carbon atoms. In electrocatalytic CO_2 reduction, the unique properties of graphene, such as its large surface area, high conductivity, and high stability, have gained significant attention. The activity of graphene is enhanced by using metallic and non-metallic doping species (such as boron, nitrogen, and phosphorus), surface tuning, and by constructing composite materials, as shown in Fig. (**8A-B**) [6].

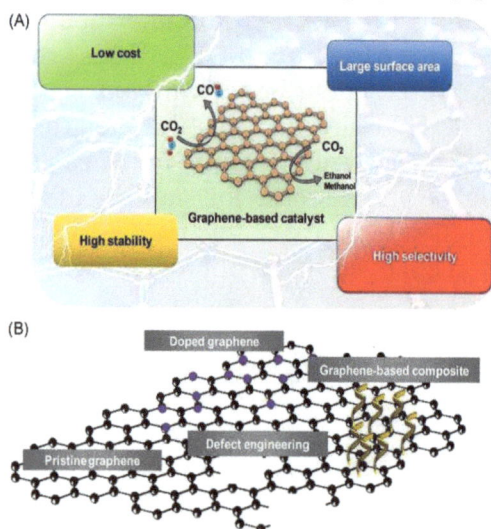

Fig. (8). (A) Advantages of graphene-based catalyst for CO_2 reduction, and (B) strategies for improving the efficiency of graphene-based catalyst for electrochemical CO_2 reduction [6].

Fig. (**9**) shows all possible reaction pathways for the synthesis of various useful products from electrochemical reduction of CO_2 [46]. The possible reactions for electrochemical CO_2 reduction are explained as the transfer of electrons to produce $CO_2^{\cdot-}$ anion radicals and use a large amount of energy to rearrange the CO_2 to $CO_2^{\cdot-}$ anion radicals [47]. Yoo *et al.* demonstrated the generation of Formic acid (HCOOH), CO, and CH_4 *via* *OCHO intermediates from CO_2

reduction [48]. The generation of CO by *COOH (carboxyl intermediates) and the transfer of electron-proton occur by the following reaction:

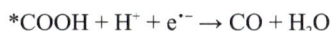

$$*COOH + H^+ + e^{\cdot-} \rightarrow CO + H_2O$$

Chai and co-workers reported that the formation of *COOH on N-doped graphene is more favorable than *OCHO based on density functional theory calculations (DFT) [47].

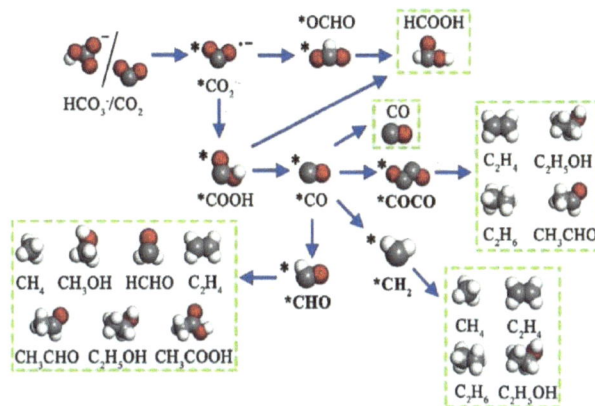

Fig. (9). Proposed reaction pathways with intermediates for different CO_2 electroreduction to fine chemicals [46].

It has been reported that *CO is a general intermediate for the generation of CH_4, HCHO, and CH_3OH. DFT calculations reveal that the Cu surface aids the generation of starting *CO intermediates thermodynamically, which results in the hydrogenation of *HCO, *H_3CO, and *H_2CO. Two products of methoxy intermediate reduction are *O and CH_4. Then CH_4 and *O are converted into H_2O. The potential barriers for C-H formation and CH_3OH generation are found to be 1.2 and 0.15 eV, respectively. The conversion of CO_2 is observed using X-ray photoelectron spectroscopy and Auger electron spectroscopy during the electrochemical measurement of working electrodes. Finally, CH_4 can be formed *via* CH_3, CH_2, and CH_1 intermediates.

Song *et al.* reported the formation of Cu nanoparticles (Cu NPs) and N-doped graphene composites. These composite materials were further studied for electrochemical CO_2 reduction. They claimed that *CO radical transformed into *OC–COH because N-doped graphene favored ethanol or n-propanol generation [50]. Zhao and co-workers demonstrated that reduced graphene oxide (RGO) was used as a support for gold NPs (~2.4 nm) [51]. This synthesized catalyst had a Faradaic efficiency (FE) ranging between 32% and 60% for the generation of CO from CO_2 reduction.

CONCLUSION

In this chapter, we have summarized the results reported in the available literature. The structure, strategies, and mechanism of the utilization of CO_2 into value-added chemicals using graphene and its composite materials are also discussed. The progress achieved by different research groups in this field demonstrates that coupling graphene with catalytic materials in a suitable manner to generate composite materials improves the efficiency of converting CO_2 into value-added chemicals. Despite significant developments obtained in this field, some fundamental and essential issues can be employed to improve catalytic efficiency. In this context, computational methods can be utilized to rationally understand the mechanism of catalytic reactions, which would guide the scientists to design more efficient and highly selective graphene-based composites to promote the conversion of CO_2 into value-added chemicals.

CONSENT FOR PUBLICATION

Not applicable.

CONFLICT OF INTEREST

The author declares no conflict of interest, financial or otherwise.

ACKNOWLEDGEMENTS

Declared none.

REFERENCES

[1]   Jacobson, M.Z. Energy Environ. Sci. 2009, 2, 148-173. (b) Tiba, S.; Omri, A. *Renew. Sustain. Energy Rev.,* **2017**, *69*, 1129-1146.

[2]   Stocker, T.F.; Qin, D.; Plattner, G.K.; Tignor, M.M.; Allen, S.K.; Boschung, J.; Nauels, A.; Xia, Y.; Bex, V.; Midgley, P.M. *Climate Change 2013: The Physical Science Basis. Contribution of Working Group 1 to the Fifth Assessment Report of the Intergovernmental Panel on Climate Change*; Cambridge University Press: Cambridge, U. K., **2014**.

[3]   Sakakura, T.; Choi, J.C.; Yasuda, H. Transformation of carbon dioxide. *Chem. Rev.,* **2007**, *107*(6), 2365-2387.
         [http://dx.doi.org/10.1021/cr068357u] [PMID: 17564481]

[4]   Sharma, N.; Dhankhar, S.S.; Kumar, S.; Kumar, T.J.D.; Nagaraja, C.M. Rational Design of a 3D Mn^{II}-Metal-Organic Framework Based on a Nonmetallated Porphyrin Linker for Selective Capture of CO_2 and One-Pot Synthesis of Styrene Carbonates. *Chemistry,* **2018**, *24*(62), 16662-16669.
         [http://dx.doi.org/10.1002/chem.201803842] [PMID: 30152564]

[5]   Sharma, N.; Dhankhar, S.S.; Nagaraja, C.M. Environment-Friendly, Co-catalyst-and Solvent-Free Fixation of CO_2 using an Ionic Zinc(II)–Porphyrin Complex Immobilized in Porous Metal-Organic Frameworks. *Sustain. Energy Fuels,* **2019**, *3*, 2977-2982.
         [http://dx.doi.org/10.1039/C9SE00282K]

[6] Hasani, A.; Teklagne, M.A.; Do, H.H.; Hong, S.H.; Van Le, Q.; Ahn, S.H.; Kim, S.Y. Graphene-Based Catalysts for Electrochemical Carbon Dioxide Reduction. *Carbon Energy,* **2020**, *2*, 158-175.
[http://dx.doi.org/10.1002/cey2.41]

[7] Zhang, Y.; Tan, Y.W.; Stormer, H.L.; Kim, P. Experimental observation of the quantum Hall effect and Berry's phase in graphene. *Nature,* **2005**, *438*(7065), 201-204.
[http://dx.doi.org/10.1038/nature04235] [PMID: 16281031]

[8] Berger, C.; Song, Z.; Li, X.; Wu, X.; Brown, N.; Naud, C.; Mayou, D.; Li, T.; Hass, J.; Marchenkov, A.N.; Conrad, E.H.; First, P.N.; de Heer, W.A. Electronic confinement and coherence in patterned epitaxial graphene. *Science,* **2006**, *312*(5777), 1191-1196.
[http://dx.doi.org/10.1126/science.1125925] [PMID: 16614173]

[9] Marques Mota, F.; Kim, D.H. From CO_2 methanation to ambitious long-chain hydrocarbons: alternative fuels paving the path to sustainability. *Chem. Soc. Rev.,* **2019**, *48*(1), 205-259.
[http://dx.doi.org/10.1039/C8CS00527C] [PMID: 30444252]

[10] Wu, T.; Lin, J.; Cheng, Y.; Tian, J.; Wang, S.; Xie, S.; Pei, Y.; Yan, S.; Qiao, M.; Xu, H.; Zong, B. Porous Graphene-Confined Fe-K as Highly Efficient Catalyst for CO_2 Direct Hydrogenation to Light Olefins. *ACS Appl. Mater. Interfaces,* **2018**, *10*(28), 23439-23443.
[http://dx.doi.org/10.1021/acsami.8b05411] [PMID: 29956535]

[11] Ning, H.; Mao, Q.; Wang, W.; Yang, Z.; Wang, X.; Zhao, Q.; Song, Y.; Wu, M. N-doped Reduced Graphene Oxide Supported Cu_2O Nanocubes as High Active Catalyst for CO_2 Electroreduction to C_2H_4. *J. Alloys Compd.,* **2019**, *785*, 7-12.
[http://dx.doi.org/10.1016/j.jallcom.2019.01.142]

[12] Ma, R.; He, L.N.; Zhou, Y.B. An Efficient and Recyclable Tetraoxo-Coordinated Zinc Catalyst for the Cycloaddition of Epoxides with Carbon Dioxide at Atmospheric Pressure. *Green Chem.,* **2016**, *18*, 226-231.
[http://dx.doi.org/10.1039/C5GC01826A]

[13] Ema, T.; Miyazaki, Y.; Shimonishi, J.; Maeda, C.; Hasegawa, J.Y. Bifunctional porphyrin catalysts for the synthesis of cyclic carbonates from epoxides and CO_2: structural optimization and mechanistic study. *J. Am. Chem. Soc.,* **2014**, *136*(43), 15270-15279.
[http://dx.doi.org/10.1021/ja507665a] [PMID: 25268908]

[14] Liu, M.; Gao, K.; Liang, L.; Sun, J.; Sheng, L.; Arai, M. Experimental and Theoretical Insights into Binary Zn-SBA-15/KI Catalysts for the Selective Coupling of CO_2 and Epoxides into Cyclic Carbonates under Mild Conditions. *Catal. Sci. Technol.,* **2016**, *6*, 6406-6416.
[http://dx.doi.org/10.1039/C6CY00725B]

[15] Jiang, X.; Gou, F.; Chen, F.; Jing, H. Cycloaddition of Epoxides and CO_2 Catalyzed by Bisimidazole-Functionalized Porphyrin Cobalt (III) Complexes. *Green Chem.,* **2016**, *18*, 3567-3576.
[http://dx.doi.org/10.1039/C6GC00370B]

[16] Whiteoak, C.J.; Nova, A.; Maseras, F.; Kleij, A.W. Merging sustainability with organocatalysis in the formation of organic carbonates by using $CO_{(2)}$ as a feedstock. *ChemSusChem,* **2012**, *5*(10), 2032-2038.
[http://dx.doi.org/10.1002/cssc.201200255] [PMID: 22945474]

[17] Monassier, A.; D'Elia, V.; Cokoja, M.; Dong, H.; Pelletier, J.D.; Basset, J.M.; Kühn, F.E. Synthesis of Cyclic Carbonates from Epoxides and CO_2 under Mild Conditions using a simple, highly Efficient Niobium-based Catalyst. *ChemCatChem,* **2013**, *5*, 1321-1324.
[http://dx.doi.org/10.1002/cctc.201200916]

[18] Song, Q.W.; He, L.N.; Wang, J.Q.; Yasuda, H.; Sakakura, T. Catalytic Fixation of CO_2 to Cyclic Carbonates by Phosphonium Chlorides Immobilized on Fluorous Polymer. *Green Chem.,* **2013**, *15*, 110-115.
[http://dx.doi.org/10.1039/C2GC36210D]

[19]   Liang, J.; Huang, Y.B.; Cao, R. Metal-Organic Frameworks and Porous Organic Polymers for Sustainable Fixation of Carbon Dioxide into Cyclic Carbonates. *Coord. Chem. Rev.,* **2019**, *378*, 32-65. [http://dx.doi.org/10.1016/j.ccr.2017.11.013]

[20]   Ding, M.; Flaig, R.W.; Jiang, H.L.; Yaghi, O.M. Carbon capture and conversion using metal-organic frameworks and MOF-based materials. *Chem. Soc. Rev.,* **2019**, *48*(10), 2783-2828. [http://dx.doi.org/10.1039/C8CS00829A] [PMID: 31032507]

[21]   Maina, J.W.; Pozo-Gonzalo, C.; Kong, L.; Schütz, J.; Hill, M.; Dumée, L.F. Metal-Organic Framework based Catalysts for CO_2 Conversion. *Mater. Horiz.,* **2017**, *4*, 345-361. [http://dx.doi.org/10.1039/C6MH00484A]

[22]   Cui, W.G.; Zhang, G.Y.; Hu, T.L.; Bu, X.H. Metal-Organic Framework-based Heterogeneous Catalysts for the Conversion of C1 Chemistry: CO, CO_2 and CH_4. *Chem. Rev.,* **2019**, *387*, 79-120.

[23]   Olajire, A.A. Synthesis Chemistry of Metal-Organic Frameworks for CO_2 Capture and Conversion for Sustainable Energy Future. *Renew. Sustain. Energy Rev.,* **2018**, *92*, 570-607. [http://dx.doi.org/10.1016/j.rser.2018.04.073]

[24]   Zhang, S.; Zhang, H.; Cao, F.; Ma, Y.; Qu, Y. Catalytic Behavior of Graphene Oxides for Converting CO_2 into Cyclic Carbonates at one Atmospheric Pressure. *ACS Sustain. Chem.& Eng.,* **2018**, *6*, 4204-4211. [http://dx.doi.org/10.1021/acssuschemeng.7b04600]

[25]   Saptal, V.B.; Sasaki, T.; Harada, K.; Nishio-Hamane, D.; Bhanage, B.M. Hybrid Amine-Functionalized Graphene Oxide as a Robust Bifunctional Catalyst for Atmospheric Pressure Fixation of Carbon Dioxide using Cyclic Carbonates. *ChemSusChem,* **2016**, *9*(6), 644-650. [http://dx.doi.org/10.1002/cssc.201501438] [PMID: 26840889]

[26]   Lan, D.H.; Chen, L.; Au, C.T.; Yin, S.F. One-Pot Synthesized Multi-Functional Graphene Oxide as a Water-Tolerant and Efficient Metal-Free Heterogeneous Catalyst for Cycloaddition Reaction. *Carbon,* **2015**, *93*, 22-31. [http://dx.doi.org/10.1016/j.carbon.2015.05.023]

[27]   Xu, J.; Xu, M.; Wu, J.; Wu, H.; Zhang, W.H.; Li, Y.X. Graphene Oxide Immobilized with Ionic Liquids: Facile Preparation and Efficient Catalysis for Solvent-free Cycloaddition of CO_2 to Propylene Carbonate. *RSC Advances,* **2015**, *5*, 7236-72368. [http://dx.doi.org/10.1039/C5RA13533H]

[28]   Zhang, W.H.; He, P.P.; Wu, S.; Xu, J.; Li, Y.; Zhang, G.; Wei, X.Y. Graphene Oxide Grafted Hydroxyl-Functionalized Ionic Liquid: A Highly Efficient Catalyst for Cycloaddition of CO_2 with Epoxides. *Appl. Catal. A,* **2016**, *509*, 111-117. [http://dx.doi.org/10.1016/j.apcata.2015.10.038]

[29]   Tong, H.; Ouyang, S.; Bi, Y.; Umezawa, N.; Oshikiri, M.; Ye, J. Nano-photocatalytic materials: possibilities and challenges. *Adv. Mater.,* **2012**, *24*(2), 229-251. [http://dx.doi.org/10.1002/adma.201102752] [PMID: 21972044]

[30]   Inoue, T.; Fujishima, A.; Konishi, S.; Honda, K. Photoelectrocatalytic Reduction of Carbon Dioxide in Aqueous Suspensions of Semiconductor Powders. *Nature,* **1979**, *277*, 637-638. [http://dx.doi.org/10.1038/277637a0]

[31]   Grills, D.C.; Fujita, E. New Directions for the Photocatalytic Reduction of CO_2: Supramolecular, $scCO_2$ or Biphasic Ionic Liquid−$scCO_2$ Systems. *J. Phys. Chem. Lett.,* **2010**, *1*, 2709-2718. [http://dx.doi.org/10.1021/jz1010237]

[32]   Roy, S.C.; Varghese, O.K.; Paulose, M.; Grimes, C.A. Toward solar fuels: photocatalytic conversion of carbon dioxide to hydrocarbons. *ACS Nano,* **2010**, *4*(3), 1259-1278. [http://dx.doi.org/10.1021/nn9015423] [PMID: 20141175]

[33]   Ong, W.J.; Tan, L.L.; Chai, S.P.; Yong, S.T.; Mohamed, A.R. Self-Assembly of Nitrogen-Doped TiO_2 with Exposed {001} facets on a Graphene Scaffold as Photo-active Hybrid Nanostructures for

Reduction of Carbon Dioxide to Methane. *Nano Res.,* **2014**, *7*, 1528-1547.
[http://dx.doi.org/10.1007/s12274-014-0514-z]

[34]   Yang, M.Q.; Xu, Y.J. Photocatalytic conversion of CO_2 over graphene-based composites: current status and future perspective. *Nanoscale Horiz.,* **2016**, *1*(3), 185-200.
[http://dx.doi.org/10.1039/C5NH00113G] [PMID: 32260621]

[35]   Yadav, R.K.; Baeg, J.O.; Oh, G.H.; Park, N.J.; Kong, K.J.; Kim, J.; Hwang, D.W.; Biswas, S.K. A photocatalyst-enzyme coupled artificial photosynthesis system for solar energy in production of formic acid from CO_2. *J. Am. Chem. Soc.,* **2012**, *134*(28), 11455-11461.
[http://dx.doi.org/10.1021/ja3009902] [PMID: 22769600]

[36]   Kumar, P.; Sain, B.; Jain, S.L. Photocatalytic Reduction of Carbon Dioxide to Methanol using a Ruthenium Trinuclear Polyazine Complex Immobilized on Graphene Oxide under Visible Light Irradiation. *J. Mater. Chem. A Mater. Energy Sustain.,* **2014**, *2*, 11246-11253.
[http://dx.doi.org/10.1039/c4ta01494d]

[37]   Wu, T.; Zhu, C.; Han, D.; Kang, Z.; Niu, L. Highly selective conversion of CO_2 to C_2H_6 on graphene modified chlorophyll Cu through multi-electron process for artificial photosynthesis. *Nanoscale,* **2019**, *11*(47), 22980-22988.
[http://dx.doi.org/10.1039/C9NR07824J] [PMID: 31769773]

[38]   Lee, S.; Kim, D.; Lee, J. Electrocatalytic Production of C3-C4 Compounds by Conversion of CO_2 on a Chloride-Induced Bi-Phasic Cu_2O-Cu Catalyst. *Angew. Chem. Int. Ed. Engl.,* **2015**, *54*(49), 14701-14705.
[http://dx.doi.org/10.1002/anie.201505730] [PMID: 26473324]

[39]   Kuss-Petermann, M.; Orazietti, M.; Neuburger, M.; Hamm, P.; Wenger, O.S. Intramolecular Light-Driven Accumulation of Reduction Equivalents by Proton-Coupled Electron Transfer. *J. Am. Chem. Soc.,* **2017**, *139*(14), 5225-5232.
[http://dx.doi.org/10.1021/jacs.7b01605] [PMID: 28362497]

[40]   Amatore, C.; Saveant, J.M. Mechanism and kinetic characteristics of the electrochemical reduction of carbon dioxide in media of low proton availability. *J. Am. Chem. Soc.,* **1981**, *103*, 5021-5023.
[http://dx.doi.org/10.1021/ja00407a008]

[41]   Lundeen, A.J.; Van Hoozer, R. Selective Catalytic Dehydration of 2-alcohols; A New Synthesis of 1-Olefins. *J. Am. Chem. Soc.,* **1963**, *85*, 2180-2181.
[http://dx.doi.org/10.1021/ja00897a041]

[42]   Lundeen, A.J.; VanHoozer, R. Selective Catalytic Dehydration. Thoria-Catalyzed Dehydration of Alcohols. *J. Org. Chem.,* **1967**, *32*, 3386-3389.
[http://dx.doi.org/10.1021/jo01286a024]

[43]   Kim, W.; Seok, T.; Choi, W. Nafion Layer-Enhanced Photosynthetic Conversion of CO_2 into Hydrocarbons on TiO_2 Nanoparticles. *Energy Environ. Sci.,* **2012**, *5*, 6066-6070.
[http://dx.doi.org/10.1039/c2ee03338k]

[44]   Liu, Y.; Zhang, Y.; Cheng, K.; Quan, X.; Fan, X.; Su, Y.; Chen, S.; Zhao, H.; Zhang, Y.; Yu, H.; Hoffmann, M.R. Selective Electrochemical Reduction of Carbon Dioxide to Ethanol on a Boron- and Nitrogen-Co-doped Nanodiamond. *Angew. Chem. Int. Ed. Engl.,* **2017**, *56*(49), 15607-15611.
[http://dx.doi.org/10.1002/anie.201706311] [PMID: 28914470]

[45]   Montoya, J.H.; Peterson, A.A.; Nørskov, J.K. Insights into C-C Coupling in CO_2 Electroreduction on Copper Electrodes. *ChemCatChem,* **2013**, *5*, 737-742.
[http://dx.doi.org/10.1002/cctc.201200564]

[46]   Ma, T.; Fan, Q.; Li, X.; Qiu, J.; Wu, T.; Sun, Z. Graphene based materials for electrochemical Co_2 reduction. *J CO_2 Uti.,* **2019**, *30*, 168-182.

[47]   Hori, Y.; Wakebe, H.; Tsukamoto, T.; Koga, O. Electrocatalytic Process of CO Selectivity in Electrochemical Reduction of CO_2 at Metal Electrodes in Aqueous Media. *Electrochim. Acta,* **1994**,

39, 1833-1839.
[http://dx.doi.org/10.1016/0013-4686(94)85172-7]

[48]   Yoo, J.S.; Christensen, R.; Vegge, T.; Nørskov, J.K.; Studt, F. Theoretical Insight into the Trends that Guide the Electrochemical Reduction of Carbon Dioxide to Formic Acid. *ChemSusChem,* **2016**, *9*(4), 358-363.
[http://dx.doi.org/10.1002/cssc.201501197] [PMID: 26663854]

[49]   Chai, G.L.; Guo, Z.X. Highly effective sites and selectivity of nitrogen-doped graphene/CNT catalysts for CO_2 electrochemical reduction. *Chem. Sci. (Camb.),* **2016**, *7*(2), 1268-1275.
[http://dx.doi.org/10.1039/C5SC03695J] [PMID: 29910883]

[50]   Song, Y.; Peng, R.; Hensley, D.K.; Bonnesen, P.V.; Liang, L.; Wu, Z. High-Selectivity Electrochemical Conversion of CO_2 to Ethanol using a Copper Nanoparticle/N-doped Graphene Electrode. *ChemistrySelect,* **2016**, *1*, 6055-6061.
[http://dx.doi.org/10.1002/slct.201601169]

[51]   Zhao, Y.; Wang, C.; Liu, Y.; MacFarlane, D.R.; Wallace, G.G. Engineering Surface Amine Modifiers of Ultrasmall Gold Nanoparticles Supported on reduced Graphene Oxide for Improved Electrochemical CO_2 Reduction. *Adv. Energy Mater.,* **2018**, *8*, 1801400.
[http://dx.doi.org/10.1002/aenm.201801400]

CHAPTER 7

Metal-Doped Graphene Materials as Electrocatalysts in Sensors

H. C. Ananda Murthy[1,*]**, Nigussie Alebachew**[1,*]**, K B Tan**[2]**, R Balachandran**[3]**, Kah-Yoong Chan**[4]** and C R Ravikumar**[5]

[1] *Department of Applied Chemistry, School of Applied Natural Science, Adama Science and Technology University, Adama, P.O. Box.1888, Ethiopia*

[2] *Department of Chemistry, Faculty of Science, Universiti Putra Malaysia, 43400 Serdang, Selangor, Malaysia*

[3] *School of Electrical Engineering and Computing, Adama Science and Technology University, PO Box 1888, Adama, Ethiopia*

[4] *Centre for Advanced Devices and Systems, Faculty of Engineering, Multimedia University, Persiaran Multimedia, 63100, Cyberjaya, Selangor, Malaysia*

[5] *Research Center, Department of Science, East-West Institute of Technology, VTU, Bengaluru 560091, India*

Abstract: A wide range of new electrocatalytic nanomaterials are introduced from time to time and are expected to continuously develop and advance the electrocatalytic sensors for different molecules. This book chapter intends to cover the latest progress and innovations in the field of metal-doped graphene (MDG) based electrocatalysts as sensors. In addition to experimental studies, electrocatalytic sensor behavior of bioactive molecules using theoretical studies has become one of the most rapidly developing fields. It is readily evident that metals combined with graphene, doped with single transition metal atom catalysts, and prepared as electrocatalytic sensors can be employed to enhance some unique properties different from the properties of bulk graphene. This book chapter encompasses preparation techniques of MDG materials as sensors for the detection of H_2, NO_2, H_2O_2, CO, CO_2, SO_2, O_2, and H_2S from experimental and theoretical perspectives. In this respect, we present a synthesis of MDG based electrocatalysts and factors affecting sensors using both experimental and theoretical studies. The most probable reason for the highest efficiency of some metals doped graphene is described from the experimental and theoretical perspectives. Finally, the conclusion describing challenges and future outlooks for the advancement of MDG materials is also given.

*Corresponding author H.C. Ananda Murthy:** Department of Applied Chemistry, School of Applied Natural Science, Adama Science and Technology University, PO Box 1888, Adama, Ethiopia; E-mails: anandkps350@gmail.com, nigussiealebachew@gmail.com

Keywords: Bioactive molecules, Electrocatalyst, First principles, Metal doped graphene, Sensor.

INTRODUCTION

After the introduction of graphene as a novel material in 2004, the research on graphene, a two-dimensional (2D) sheet of carbon atoms consisting of carbon in a honeycomb pattern to form a single layer of atoms, is an emerging field on the horizons of material science and nanotechnology [1]. For one thing, the utilization of this 2D material shows remarkably great crystal and electronic quality, as well as potential uses [2, 3]. With this in mind, doped graphene nanocomposites for gas sensors, biosensors, and hydrogen storage energy sectors have become a popular research area.

Recently, it has been reported that due to its unique physicochemical characteristics, such as high electronic conductivity, a zero-gap semiconductor with tunable conductance and charge carrier density (200 Sm^{-1}) [3, 4], large specific surface area (2630 $m^2\ g^{-1}$) [5], a theoretical surface area of 2630 m^2g^{-1}, mobility of 200,000 $cm^2\ V^{-1}\ s^{-1}$ at a carrier density of ~1012 cm^{-2}, and the highest electrical conductivity of 106 Sm^{-1} at room temperature [6], excellent thermal stability [7], and superior mechanical strength, many research efforts are still in progress to explore these characteristics on this popular material.

However, due to the strong sp^2 linkage between the carbon atoms in the graphene plane, the easy aggregation of 2D graphene nanosheets and its irreversible phenomenon are attributed to the strong π-π stacking and Vander Waals interactions between the graphene nanosheets which suffers from diversified topological defects during its growth [8]. The major limitation is the relatively poor interaction between graphene and the metal surface, which in most cases leads to severe modification of the electronic structure of graphene, to the more concrete issues, like the preparing of metal doped graphene (MDG) materials on the surface of transition metals [9].

A serious effort has been made to tackle this challenge. An effective way is to apply a graphene material to synthesize less aggregated graphene, such as MDG composites used in the electrochemical (bio) sensors. Despite a number of articles on the interactions of detecting small molecules on the metals-doped graphene sheets, a comprehensive review on the experimental and theoretical studies on MDG based nanomaterials for electrocatalytic sensors is still rare. Hence, in the present review, our main objective is to summarize the perspective and current challenges of MDG based nanomaterials for electrocatalyst sensors from experimental and theoretical calculations. This book chapter tries to make links between experimental studies and current density functional theory (DFT) calculations for electrocatalyst sensing of small molecules, including reduction of oxygen and hydrogen, H_2 detection, sensing harmful gas molecules. Finally, the challenges and the trend of future efforts associated with high-performance electrocatalytic sensors based on MDG for a more penetrating analysis of small molecules are proposed.

SYNTHESIS OF MDG ELECTROCATALYSTS

As mentioned in the above section, as a result of many exceptional properties of graphene and its related materials such as graphene oxide (GO), a substantial amount of research has been carried out applying these materials directly or in combination with other nanomaterials such as metals nanoparticles. During the synthesis of MDG materials, the formation of the structural defects on graphene provides more operating sites for further alteration with different practical groups promoting the selective electrocatalysts. In the following section, we will describe the assembly of MDG for the electrocatalysts in sensor application.

Evaluating the process of the synthesis of graphene and its derivatives, it is important to note that in some simple top-down synthesis, the desired structure and properties are mainly contingent upon the size, shape, and functional groups attached to the surface of the material. Several synthetic routes have been reported through which MDG materials can be prepared for fabricating hybrid MDGs for electrocatalyst sensor applications. In the context of testing various methods, it is imperative to note the doping of transition metals by using the simple exfoliation of the bulk graphite method. For the sake of brevity, doping of transition metals (Mn, Fe, Co, Ni) uranium (U) and thorium (Th) [10, 11], accumulation of platinum (Pt) nanoparticles on the surface of graphene nanosheet using a microwave-assisted method [12] are among the methods in which the metal nanoparticles stuck/ adhered to the surface of the graphene sheet. In some studies, transition metal-doped graphene hybrids with new properties were made by exfoliation of metal-doped graphite oxide precursors in either hydrogen or nitrogen atmospheres. Classifications of such materials suggest that the H_2 exfoliated materials had higher C/O ratios [10]. With a similar notion, uranium graphene and thorium graphene hybrids were formed by the reaction of UO_2^{2+} and Th^{4+} ions with graphene oxide and subsequent exfoliation in hydrogen and nitrogen (H_2 and N_2) atmospheres. The characterization results revealed high C/O ratios, which confirm the formation of highly reduced graphene hybrids, predominantly in a hydrogen atmosphere [11]. In a number of experiments, it was shown that graphene layers of very high-quality iron-doped thermally reduced graphene material were used as a magnetic modifier of a carbon-based electrode [13].

The findings of the amazing transport properties of graphene renewed interest in the examination of the metal/graphene interfaces. Here we present some studies ranging from the basic problems, such as the right description of the comparatively weak interaction between graphene and metal surfaces. This, in most cases, leads to the severe alteration of the electronic structure of graphene to the more useful issues like the organization of the ordered arrays of clusters on the graphene, particularly on the surface of 4d and 5d metals. This can be used as an ideal system for the studying of catalytic characteristics of MDG materials. In spite of several excellent studies which focussed on the modification of MDG based electrocatalysts and their applications in solving environmental issues, however, an extensive review by Hidalgo-Manrique *et al.* showed the versatile properties of copper doped graphene composites synthesized by extensive mixing techniques like mechanical stirring, magnetic stirring, sonication and vortex mixing, electrochemical deposition, and chemical vapour deposition methods. According to the reviewers, in addition to the synthesis method, the mechanical properties of copper doped graphene mixtures are highly dependent on many factors. Some of these factors include graphene nanosheet manufacturing conditions, graphene structural defects, the effect of graphene modification when decorating graphene with nickel, and ductility, which is reduced when a large amount of graphene is added to the composite. Moreover, the microstructure properties of copper doped graphene composites, as well as the electronic and optical properties, are affected by controlling the above external factors rather than changing synthetic routes [14].

Even though much efforts have been devoted to synthesizing MDG materials, considerable challenges are still existing to attain single sheets of graphene from bulk graphite. To solve this problem, many researchers preferred the mechanical exfoliation technique. However, to have defect-free graphene, there have been some difficulties like shape, size, edge, and layers of graphene due to random exfoliation [15]. To prepare single and multiple-layer graphene in large quantities or functionalized graphene that can be utilized for the making of these composites, others applied direct sonication and dissolution methods (liquid phase exfoliation methods) [16]. Therefore, for the production of MDG nanocomposites, which generally requires large quantities of homogeneously distributed metals such as transition metals and graphene sheets, the top-down approach, chemical and/or thermal reduction of graphite derivative such as graphite oxide as precursor using exfoliation technique is used for the electrocatalyst sensor applications [10, 13].

Despite the application of different types of precursors, modifications are conducted to enhance the properties of MDG composites in a desirable way. However, electrocatalysts' sensor active sites of graphene induced by metals are affected by the nitrogen and hydrogen atmospheres during the thermal exfoliation preparation technique. The thermal reduction of metal doped-graphite oxide starting materials was carried out in nitrogen and hydrogen atmospheres. The outcomes of these two atmospheres as well as the metal elements on the features and catalytic abilities of the cross materials prepared from noble metals (Pd, Ru, Rh, Pt, Au, Ag) doped graphene hybrids for electrocatalysts were studied. The study shows that the noble metal-doped graphene exfoliated in a nitrogen atmosphere contained more oxygen-containing groups and a lower density of faults on their surfaces as compared to hybrids exfoliated in a hydrogen atmosphere. In the study, the thermal exfoliation method was employed to synthesize uranium doped graphene oxide (labeled U-GO) and thorium doped graphene (Th-GO) electrocatalysts for sensor devices. The doped graphene oxide slurries, U-GO and Th-GO, were later thermally exfoliated and decreased in a hydrogen or nitrogen atmosphere for the development of different graphene-based nanohybrid electrocatalysts. A marginally acidified aqueous solution of thorium nitrate (Th $(NO_3)_4$) and uranium nitrate $UO_2(NO_3)_2$ were used as precursors (Table **1**). The general applications of graphene materials are depicted in Fig. (**1**).

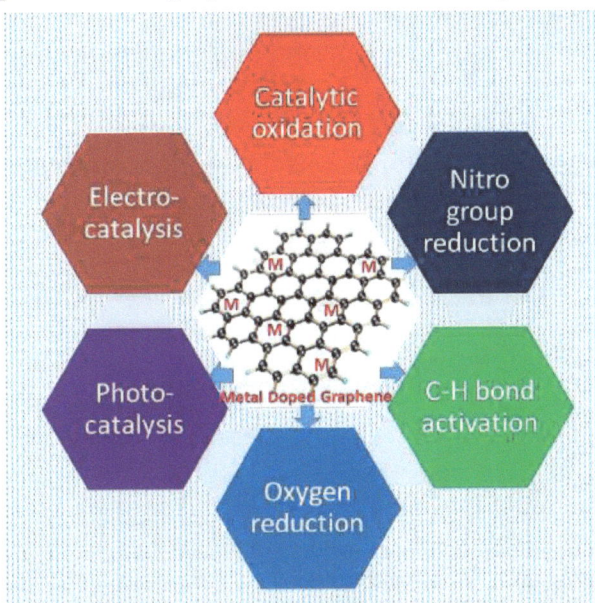

Fig. (1). General applications of metal-doped graphene material [10].

Though the hybrid materials showed unique electrochemical properties as opposed to the undecorated graphene materials, evidence from the characterization of the materials confirms that the exfoliated materials have a strong impact on the quantity of oxygen-containing groups as well as the density of faults on the surface of the hybrid materials. By the same token, components exfoliated in a nitrogen N_2 atmosphere containing more oxygen-containing groups and having a lower density of faults than those exfoliated in hydrogen H_2 [10, 11, 17].

MDG Materials as Sensors

Generally, sensors are comprised of two basic elements: a receptor and a transducer. The receptor can be of any organic or inorganic material that works with a specific analyte or family of analytes. The transducer, however, is an element that changes the recognition that happens between the analyte and the receptor into a quantifiable signal. This signal can appear in many forms including, electrical, electrochemical, and optical. In this section, we will focus on works completed by different research groups on the growth of MDG electrocatalysts as sensors for the discovery of chemical and biological analytes.

Reduction of Oxygen and Hydrogen

In MDG materials, metal performs the role of the catalytic center, whereas graphene is the high area conductor. In the U-GO and Th-GO materials, graphene oxide slurries were later thermally exfoliated and decreased in a hydrogen or nitrogen atmosphere for the development of different graphene-based nanohybrid electrocatalysts. Because of the larger sorption capacity of graphene oxide toward uranyl ion as compared to thorium ions and a higher concentration of uranyl ions, the electrocatalytic activity toward decreasing oxygen and hydrogen peroxide shows better electrocatalysts for sensor devices.

One of the key steps in the electrolytic sensors is the ability of electron transfer between the analytes and to the electrodes completing the circuit through the developed material without any interference. To attain uninterrupted electron transfer and to rise above all other constraints, many researchers have developed different types of noble MDG hybrid materials which can be used for direct sensing of hydrogen peroxide (H_2O_2), a tough oxidizing agent; oxidation of hydrazine (N_2H_4), a reducing agent; and reduction of oxygen (O_2) dissolved in potassium hydroxide (KOH). The reduction peaks of H_2O_2 on the exterior of noble metal-doped graphene and control materials, such as graphene without any metals (G/N_2 and G/H_2), exhibit different situations. In this regard, the characteristics and catalytic properties of the noble metal nanoparticles supported on graphene materials show an enhancement in electron transfer rate and catalyze reduction and oxidation responses of hydrogen peroxide, hydrazine, and oxygen in alkaline materials, which can be used as electrochemical sensing and biosensing as well as electrocatalysts. The G/N_2 and G/H_2 materials had extreme reduction peak potential values of -892 and -784 mV, which confirmed the absence of hydrogen peroxide, hydrazine, and oxygen on the surfaces of G/N_2 and G/H_2 materials [17]. A similar conclusion was reported in the previous study [18].

It has been revealed from the literature review that graphene-based nanocomposites are very well established as electrocatalysts. With a similar notion, a promising alternative of important strategy for inert metals affected by the oxidation states and restricted by low availability and high expenses has been reported using transition metal-doped graphene mixtures for electrocatalysts application. The transition metal (Mn, Fe, Co, Ni) nanoparticles doped on graphene hybrids of different oxidation states under different exfoliation atmospheres were compared in terms of electrocatalytic activity. The authors reported that graphene hybrid materials exfoliated in a nitrogen atmosphere exhibited better performance as compared to those exfoliated in a hydrogen atmosphere. In this regard, nickel-doped graphene mixtures displayed superior electrocatalytic activities towards the decrease of O_2 as compared to plain graphene. This attributed to the fact that the occurrence of dopant transition metals had a substantial/important effect on the electrocatalytic activity of the hybrid materials. According to their findings, due to differentiation/distinction in the oxidation states among transition metal nanoparticles and different interactions among these metals with graphene, exfoliation in the H_2 atmosphere led to the decrease of oxidation states of the transition metal nanoparticles. This led to metals with lesser oxidation states acting as electron acceptors and may minimize chelation to the graphene supports [10].

H_2 Detection

Several works reported in the literature have explored the use of graphene-based nanomaterials as electrocatalysts for degradation of pollutants. It was reported that the heterojunction at the connection interface forms if the type of charge in the graphene base channel is different than the type of charge produced in the graphene below the contacts. Hence, the kind of charge and average doping is controlled by the kind of metal used for the connection. Graphene-metal connection regions exhibit distinctive traits as compared to the conservative semiconductor-metal connection. Graphene can be doped by the connecting metal, and the work function depends intensely on the metal varieties. Work function joining can create a p-n junction in the graphene channel, causing asymmetrical conductance performance and contact resistance [19]. Similarly, the heterojunction at the graphene connection interface can be reverse modulated by contact with molecular hydrogen (H_2). The investigation of this material leads to the choice of new graphene-based electrocatalysts for the selective discovery of H_2, among various gases, as well as helium, nitrogen, argon, and oxygen. It is noteworthy that the DFT calculations determine the N-edged faults in graphene, which can heighten the binding energy of metals. This is because the problem with N-substitution is electron-deficiency, which can stop the metals from gathering on graphene sheets [20]. It has been reported that the most usual method to heighten the adsorption energy of H_2 on carbon substances is the insertion of defects or heteroatoms. This is due to the high binding energy for metal atoms on carbon materials, increasing total mass of the host materials, and the extreme connection between transition metals and H_2, which highly affect H_2 adsorption [21].

Table 1. Typical synthetic method, the type of precursors used for the preparation of MDG and their related applications.

Materials	Synthetic routes	Precursors and others	Applications	Ref
Mn, Fe, Co, Ni-doped Graphene	Thermal exfoliation (nitrogen or hydrogen atmosphere)	Graphite oxide, manganese (II) acetate tetrahydrate($C_4H_{14}MnO_8$), iron (III) nitrate nonahydrate ($FeH_{18}N_3O_{18}$), cobalt (II) nitrate hexahydrate (Co $(NO_3)_2 \cdot 6H_2O$), and nickel (II) nitrate hexahydrate (Ni $(NO_3)_2 \cdot 6H_2O$)	reduction of O_2	[10]
Pt doped Graphene	Microwave-assisted method	Graphite oxides, $H_2PtCl_6 \cdot nH_2O$ n =5.5	hydrogen evolution	[12]
Iron-doped Graphene	Thermal exfoliation	Graphite oxides, Fe $(NO_3)_3$ $.9H_2O$	current sensing	[13]
Uranium- and Thorium-Doped Graphene	Thermal exfoliation (nitrogen or hydrogen atmosphere)	Graphene oxide, thorium nitrate, $UO_2(NO_3)_2$)	reduction of oxygen and hydrogen peroxide	[11]
Pd, Ru, Rh, Pt, Au, Ag doped graphene	Thermal exfoliation in nitrogen (G-M/N$_2$) or hydrogen (G-M/H$_2$)	graphite oxide, palladium (II) chloride, ruthenium (III) chloride, rhodium (III) chloride, ammonium hexa-chloroplatinate (IV), gold (III) chloride, and silver nitrate.	electrochemical sensing and biosensing as well as electrocatalysis	[17]
Pd Nanoparticles Decorated N-Doped Graphene Quantum Dots@ N-Doped Carbon Hollow Nanospheres	Bottom-up method	K_2PdCl_4, SiO_2 nanospheres	detection of H_2O_2 secreted from living cancer cells	[22]

Detecting Dangerous Gas Molecules CO, CO$_2$, SO$_2$, and H$_2$S

As described in the previous section, in cases of electrocatalysts, graphene-supported metal nanostructured substances have shown promising uses because of some innovative advantages. As an example, the huge surface area and 2D flexibility of graphene nanosheets can provide enough space to contain various metallic nanomaterials and also prevent accumulation. In addition, the electrical conductivity of graphene nanosheets advances the electron transfer rate on the outside.

STUDY USING FIRST PRINCIPLES

The field of research and development of MDG electrocatalysts as sensors is currently undergoing an exciting development with increasing achievements. In parallel with the experimental methods, there are lots of theoretical studies using DFT study with different basis sets (Table **2**) focusing on MDG materials toward sensing materials such as NO on Pt-decorated graphene [23], palladium doped graphene [24], oxygen decreasing reaction over the Pt$_4$ and Pt$_3$M(M = Fe, V) groups on the O-doped graphene substrate [25], Al-doped graphene as altered nanostructure sensor for diethyl ether (DEE), ethyl methyl ether (EME), dimethyl ether (DME) [26], adsorptions of CO on pristine, Fe⁻, Ru⁻, Os⁻, Co⁻, Rh⁻, Ir⁻, Ni⁻, Pd⁻, and Pt-doped graphene [27]. Moreover, a theory-based study has found the effect of defects on graphene.

Due to higher binding energy and lower bond distance of Stone - Wales defect in graphene, heightened adsorption of H$_2$CO has been shown than with ideal or single-walled (SW) defected graphene [28].

A promising investigation using DFT calculation could be used to compare the adsorption of three distinctive oil-dissolved gases in transformers (C$_2$H$_2$, CH$_4$, and CO) on immaculate graphene, and Mn-doped graphene (Mn-graphene) to achieve a perfect gas sensing material (Fig. **2**). The theoretical calculation shows weak adsorption affinity of the pristine graphene surface to the gases. According to the authors, except for CH$_4$, the adsorption energy of the C$_2$H$_2$ and CO gases on Mn-graphene has mainly increased due to the electronic hybridization (week adsorption and small conductivity change). Mn-doped graphene was applied to study the sensitivity and selectivity of three distinctive oil-dissolved gases in transformers: C$_2$H$_2$, CH$_4$, and CO [26]. From (Fig. **2**), it is possible to examine the DOS *vs.* energy curves prior to and post Mn doping plot. It reveals the enhancement of conductivity of the system as the electronic conditions near the Fermi level grow due to the Mn dopant atom.

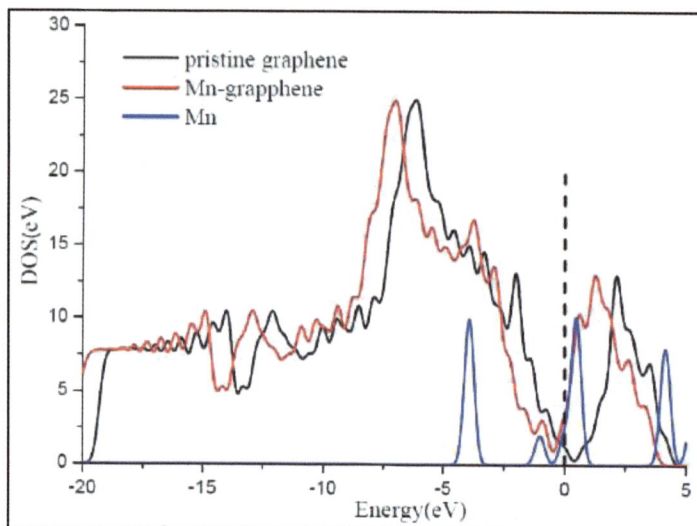

Fig (2). The DOS of pristine and Mn-doped graphene. The dashed line represents the Fermi level. (Adapted from Adsorption of C_2H_2, CH_4 and CO on Mn-doped graphene: Atomic, electronic, and gas-sensing properties [26].

The DOS distributions for Mn-graphene structure and CH_4/Mn-graphene structure show little charge transfer during adsorption. The two DOS curves are mainly overlapped specifically around the Fermi level. Both Mn-graphene system and CH_4/Mn-graphene exhibit good conductivity. An increase in DOS near -5.5 eV and -13 eV is observed due to the contribution of CH4 adsorption. Overlapping sections between the orbital paths of Mn-4*d* and H- can be found. Moreover, only a single peak of DOS in adsorbed CH_4 just places at -13 eV. Generally, from these points of view, it has been indicated that both immaculate and Mn-doped graphene exhibit weak gas sensing characteristics to CH_4 gas. Moreover, the post-DOS from the C-2*p* and Mn-3*d* distributions still show some overlapping range, but the curves decrease near -1 eV. Furthermore, the rise in valence band range from -1 eV to -2.5 eV excites electron conversion from the valence band to the conduction band. Hence the conductivity of material rises after C_2H_2 adsorb on Mn-graphene exterior; in addition, the peak value of DOS curves mainly places below Fermi level, and its peak value falls after C_2H_2 adsorption, respectively. This demonstrates that the distinctive DOS alters during the adsorption process, implying that Mn-graphene might be a potential C_2H_2 gas sensing substance [26].

A different situation is observed from the DOS distributions of C_2H_2 and CH_4. Hence, the DOS dramatically increases from -2.5 eV to Fermi level and falls from Fermi level to 1.2 eV. This shows that more electrons can change from valence to conduction band, causing an increase in conductivity. It can be seen that the overlap of interaction between Mn-4d and C-2p was evident at -6 eV, -1.5 eV, and the Fermi level, which demonstrates the strong chemical relationship between the C atom and Mn atom. The DOS peak of C-2p around -6 eV shows the appearance of the novel peak in the DOS, and the powerful chemical relationship between the C atom and Mn atom makes the adsorption mechanism possible. Finally, three peaks of DOS occur near - 5 eV, - 2.8 eV, and the Fermi level. Mn-graphene exhibits good adsorption and gas-sensing characteristics to the CO gas molecule, which is crucial for discovering the oil-dissolved gases in transformers [26].

A DFT examination of CO adsorption on VIIIB transition metal (TM) doped graphene sheets has been reported by pointing the graphene sheet C atoms to the adsorption sheet site and by directing the O atom towards the adsorption spot. The adsorption energies of CO adsorbed on the TM-doped graphene sheet by directing its C atoms toward the sheet adsorption spot were reported to be between -21.13 and -41.62 kcal/mol. Their comparative magnitudes were in the following order: CO/Os-graphene (-41.62 kcal/mol) > CO/Ir-graphene (-36.30 kcal/mol) > CO/Fe-graphene (-33.77 kcal/mol) > CO/Pt-graphene (-29.87 kcal/mol) > CO/Ru-graphene (-28.07 kcal/mol) > CO/ Ni-graphene (-23.61 kcal/mol) > CO/Rh-graphene (-23.35 kcal/mol) > CO/Co-graphene (-21.74 kcal/ mol) > CO/Pd-graphene (-21.13 kcal/mol). The adsorption energies of CO adsorbed on the TM-doped graphene sheet by directing its O atom towards the sheet adsorption spot were between -4.61 and -12.60 kcal/mol. Their comparative magnitudes were reported in the order: CO/Fe-graphene (-12.60 kcal/mol) > CO/Os-graphene (-10.96 kcal/mol) > CO/Co-graphene (-8.54 kcal/mol) >CO/ Ru-graphene (-8.42 kcal/mol) >CO/Ni-graphene (-7.25 kcal/mol) > CO/Pd-graphene (-5.67 kcal/ mol) > CO/Rh-graphene (-5.52 kcal/mol) >CO/Ir-graphene (-5.31 kcal/mol) >CO/Pt-graphene (-4.61 kcal/mol). Surprisingly, Os-doped graphene showed the highest interaction with the CO molecule when directing its C atom toward the adsorption spot. In addition to this, the Fe-doped graphene showed the highest interaction with the CO molecule when directing its O atom towards the sheet adsorption spot. Also, the adsorption strengths of CO adsorbed on the graphene sheet when directing its C atom toward the sheet adsorption spot were greater when directing the O atom toward the sheet adsorption spot. On the other hand, immaculate graphene was impervious to CO molecules. According to the authors, the atom of the CO molecule showed a higher interaction with TM-doped graphene sheets as compared to the O atom [27].

It should be obvious that improvement in the field of computational chemistry raised the standard on graphene materials. From the point of view of modern computational chemistry, the most striking feature of the works of high development in the DFT is the effect of the size of metals doped on graphene for various applications. A study was conducted to explore the size effect for palladium (Pd) doped single-vacancy defective graphene (SDG) exterior to the adsorption of AsH_3 and its dehydrogenated products on Pd using DFT calculations. The study shows that Pd_6 nanocluster linked strongly to the SDG surface, while adsorption of AsH_x (x = 0 – 3) on the most stable Pd_n doped SDG revealed dehydrogenated arsine mixtures adsorbed onto the surface more sturdily than the pristine AsH_3 molecule.

Computational chemistry has, of course, become vastly more mathematical over time. Nevertheless, the development of electrocatalysts sensors was archaic in style, even from its own time. But due to the advancement of nanotechnology, the implementation of computational chemistry on graphene for sensor purposes needs to be considered in this review. Even though experimental studies disclosed the adsorption of some toxic gases on metal-doped graphene materials, it should be obvious that computational calculations are becoming one of the sources of information for experimental studies. For environmental pollution study, more Fe-modified graphene (FeG) nanosheets towards dangerous gas toxins (CO, CO_2, SO_2 and H_2S) show that these gas toxins can be chemisorbed onto FeG with adsorption energies of 0.54 to 1.80 eV, which is 390% stronger than the adsorption energies to those onto inherent graphene in the range of 0.08 to 0.28 eV. When CO, CO_2, SO_2 and H_2S particles are adsorbed onto FeG, the charge transfer processes cause alteration in the conductance of FeG. A detailed DFT study shows that the HOMO-LUMO gap of the FeG system under gas adsorption is estimated to be useful for sensing uses. In other words, adsorption of these gas pollutants on FeG surface disclosing the effect of O_2 on an aerobic environment for the sensing property of FeG is dependent on the HOMO-LUMO energy gap. In this regard, O_2 molecules act as electron acceptors from the FeG. This creates holes that cause charge transfer, decreasing the conductance of FeG. Therefore, during the capture of gas particles onto FeG in the absence of O_2, the electronic constitution of FeG is faintly sensitive to the adsorption of SO_2, and it could be determined as an ineffective material for SO_2 sensing. On the other side, the CO_2 adsorption becomes unstable when O_2 is present, while the reaction of the electronic properties of FeG towards H_2S is obstructed when O_2 is present [29].

Table 2. Typical theoretical method and the type of metal used for the preparation of metal-doped graphene supported electrolytic sensors and their related applications.

Materials	Theoretical method	Applications	Ref
Ag, Au, Ca, Li, Mg, Pd, Pt, Sc, Sr, Ti, Y, and Zr/Graphene	DFT	H_2 adsorption	[21]
Fe-doped Graphene	DFT (PBE/Def2-SVP level)	adsorption and sensing of CO, CO_2, SO_2 and H_2S	[30]
Pd doped Graphene	DFT	adsorption of AsH_3	[24]
Li, Na, and Ca doping Graphene	DFT(DFT-D2)	H_2 adsorption	[31]

(Table 2) cont.....

Materials	Theoretical method	Applications	Ref
Al-, Si-, Mn-, Fe-, Co-, Ni-, Pd-, Ag-, Pt-, and Au/ Graphene	DFT (DMol³ code)	oxygen reduction reaction (ORR)	[32]
Fe-, Ru-, Os-, Co-, Rh-, Ir-, Ni-, Pd-, and Pt-doped Graphene were	DFT (B3LYP/LanL2DZ)	CO adsorption	[27]
Pt-decorated Graphene	DFT (6-31G (d, p) basis set by applying BLYP density functional)	adsorbent/gas sensors for NO	[23]
Al-doped Graphene	DFT(level of B3LYP/6–31G(d,p))	sensor for Diethyl ether (DEE), Ethyl methyl ether (EME), and Dimethyl ether (DME)	[26]
Cr, Mn, and Co-doped Stone–Wales defected graphene	DFT (Dmol³ code)	sensor for H_2CO molecule	[28]
A DFT Study of Arsine Adsorption on Palladium Doped Graphene: Effects of Palladium Cluster Size	DFT	adsorption of AsH_3 and	[33]
Fe-doped graphene nanosheet	DFT	adsorption platform of harmful gas molecules (CO, CO_2, SO_2, and H_2S), and the co-adsorption in O_2 environments	[29]

CONCLUSION

Despite substantial progress being made in the design and application of metal-doped graphene hybrid nanomaterials for electrocatalytic sensing of small molecules, the development of novel methods and techniques in the synthesis of MDG nanomaterials with new structures and extraordinary activities is still imperative. More importantly, an in-depth understanding of the structure-property relationship of MDG based electrocatalyst and exploring MDG using experimental and theoretical calculations for successfully applying them as electrocatalyst sensors still needs thorough investigation. Insufficient work has been undertaken to understand the link between the metal-doped graphene chemistry and the structure of graphene defects sheets. In this context, our book chapter presents results that establish a useful framework for graphene-based electrolyte sensors. Hence, further progress in this endeavor will need novel manufacturing methods which allow control over uniformed di-vacancy structures in graphene-based materials. More research could be undertaken to measure the effects of other exfoliation/reduction techniques and the performances and size of other metals, such as base metals held on graphene substances to be produced at a large scale for clinical diagnostic, food safety, pharmaceutical development, environmental monitoring will progressively appear in the near future.

CONSENT FOR PUBLICATION

Not applicable.

CONFLICT OF INTEREST

The author declares no conflict of interest, financial or otherwise.

ACKNOWLEDGEMENTS

The authors are grateful for the support rendered by Adama Science and Technology University, Ethiopia.

REFERENCES

[1]   Ortiz Balbuena, J.; Tutor de Ureta, P.; Rivera Ruiz, E.; Mellor Pita, S. Enfermedad de Vogt-Koyanag--Harada. *Med. Clin. (Barc.)*, **2016**, *146*(2), 93-94.
[http://dx.doi.org/10.1016/j.medcli.2015.04.005] [PMID: 26004271]

[2]   Bayisa, M.C.; Ananda Murthy, H.C. Effect of reinforcement of reduced graphene oxide on Mechanical Properties of Concrete nanocomposite. *J. Mater. Environ. Sci.*, **2020**, *11*(6), 844-855.

[3]   Ananda Murthy, H.C.; Kiflom, G.K.; Ravikumar, C.R. Nagaswarupa, H. P.; Aschalew, T; Tegene, D.; Graphene-supported nanomaterials as electrochemical sensors: A mini review. *Results in Chemistry*, **2021**, *3*, 100131.
[http://dx.doi.org/10.1016/j.rechem.2021.100131]

[4]   Stankovich, S.; Dikin, D.A.; Piner, R.D.; Kohlhaas, K.A.; Kleinhammes, A.; Jia, Y.; Wu, Y.; Nguyen, S.T.; Ruoff, R.S. Synthesis of Graphene-Based Nanosheets *via* Chemical Reduction of Exfoliated Graphite Oxide. *Carbon*, **2007**, *45*(7), 1558-1565.
[http://dx.doi.org/10.1016/j.carbon.2007.02.034]

[5]   Chang, C.W.; Liao, Y.C. Accelerated Sedimentation Velocity Assessment for Nanowires Stabilized in a Non-Newtonian Fluid. *Langmuir*, **2016**, *32*(51), 13620-13626.
[http://dx.doi.org/10.1021/acs.langmuir.6b03602] [PMID: 27976911]

[6]   Dreyer, D.R.; Park, S.; Bielawski, C.W.; Ruoff, R.S. The chemistry of graphene oxide. *Chem. Soc. Rev.*, **2010**, *39*(1), 228-240.
[http://dx.doi.org/10.1039/B917103G] [PMID: 20023850]

[7]   Zhu, Y.; Murali, S.; Cai, W.; Li, X.; Suk, J.W.; Potts, J.R.; Ruoff, R.S. Graphene and graphene oxide: synthesis, properties, and applications. *Adv. Mater.*, **2010**, *22*(35), 3906-3924.
[http://dx.doi.org/10.1002/adma.201001068] [PMID: 20706983]

[8]   Bae, S.; Kim, H.; Lee, Y.; Xu, X.; Park, J.S.; Zheng, Y.; Balakrishnan, J.; Lei, T.; Kim, H.R.; Song, Y.I.; Kim, Y.J.; Kim, K.S.; Ozyilmaz, B.; Ahn, J.H.; Hong, B.H.; Iijima, S. Roll-to-roll production of 30-inch graphene films for transparent electrodes. *Nat. Nanotechnol.*, **2010**, *5*(8), 574-578.
[http://dx.doi.org/10.1038/nnano.2010.132] [PMID: 20562870]

[9]   Dedkov, Y.; Voloshina, E. Graphene Growth and Properties on Metal Substrates. In: *Journal of Physics Condensed Matter*; IOP Publishing , **2015**; p. 303002.
[http://dx.doi.org/10.1088/0953-8984/27/30/303002]

[10]   Toh, R.J.; Poh, H.L.; Sofer, Z.; Pumera, M. Transition metal (Mn, Fe, Co, Ni)-doped graphene hybrids for electrocatalysis. *Chem. Asian J.*, **2013**, *8*(6), 1295-1300.
[http://dx.doi.org/10.1002/asia.201300068] [PMID: 23495248]

[11]   Sofer, Z.; Jankovský, O.; Šimek, P.; Klímová, K.; Macková, A.; Pumera, M. Uranium- and thorium-

doped graphene for efficient oxygen and hydrogen peroxide reduction. *ACS Nano,* **2014**, *8*(7), 7106-7114.
[http://dx.doi.org/10.1021/nn502026k] [PMID: 24979344]

[12]   Ullah, K.; Ye, S.; Zhu, L.; Jo, S.B.; Jang, W.K.; Cho, K.Y.; Oh, W.C. Noble Metal Doped Graphene Nanocomposites and Its Study of Photocatalytic Hydrogen Evolution. *Solid State Sci.,* **2014**, *31*, 91-98.
[http://dx.doi.org/10.1016/j.solidstatesciences.2014.03.006]

[13]   Lim, C.S.; Ambrosi, A.; Sofer, Z.; Pumera, M. Magnetic control of electrochemical processes at electrode surface using iron-rich graphene materials with dual functionality. *Nanoscale,* **2014**, *6*(13), 7391-7396.
[http://dx.doi.org/10.1039/C4NR01985G] [PMID: 24873903]

[14]   Hidalgo-manrique, P.; Lei, X.; Xu, R.; Zhou, M.; Kinloch, I.A.; Young, R.J. Copper / Graphene Composites : A Review. *J. Mater. Sci.,* **2019**, *54*(19), 12236-12289.
[http://dx.doi.org/10.1007/s10853-019-03703-5]

[15]   Cai, J.; Ruffieux, P.; Jaafar, R.; Bieri, M.; Braun, T.; Blankenburg, S.; Muoth, M.; Seitsonen, A.P.; Saleh, M.; Feng, X.; Müllen, K.; Fasel, R. Atomically precise bottom-up fabrication of graphene nanoribbons. *Nature,* **2010**, *466*(7305), 470-473.
[http://dx.doi.org/10.1038/nature09211] [PMID: 20651687]

[16]   Kim, H.; Abdala, A.A.; Macosko, C.W. Graphene/Polymer Nanocomposites. *Macromolecules,* **2010**, *43*(16), 6515-6530.
[http://dx.doi.org/10.1021/ma100572e]

[17]   Giovanni, M.; Poh, H.L.; Ambrosi, A.; Zhao, G.; Sofer, Z.; Šaněk, F.; Khezri, B.; Webster, R.D.; Pumera, M. Noble metal (Pd, Ru, Rh, Pt, Au, Ag) doped graphene hybrids for electrocatalysis. *Nanoscale,* **2012**, *4*(16), 5002-5008.
[http://dx.doi.org/10.1039/c2nr31077e] [PMID: 22763466]

[18]   Liang, Y.; Wang, H.; Zhou, J.; Li, Y.; Wang, J.; Regier, T.; Dai, H. Covalent hybrid of spinel manganese-cobalt oxide and graphene as advanced oxygen reduction electrocatalysts. *J. Am. Chem. Soc.,* **2012**, *134*(7), 3517-3523.
[http://dx.doi.org/10.1021/ja210924t] [PMID: 22280461]

[19]   Song, S. M.; Cho, B. J. Contact Resistance in Graphene Channel Transistors. *Carbon letters ,* **2013**, *14*(3), 162-170.
[http://dx.doi.org/10.5714/CL.2013.14.3.162]

[20]   Cadore, A.R.; Mania, E.; De Morais, E.A.; Watanabe, K.; Taniguchi, T.; Lacerda, R.G.; Campos, L.C. Metal-Graphene Heterojunction Modulation *via* H_2 Interaction. *Appl. Phys. Lett.,* **2016**, *109*(3)
[http://dx.doi.org/10.1063/1.4959560]

[21]   Fair, K.M.; Cui, X.Y.; Li, L.; Shieh, C.C.; Zheng, R.K.; Liu, Z.W.; Delley, B.; Ford, M.J.; Ringer, S.P.; Stampfl, C. Hydrogen Adsorption Capacity of Adatoms on Double Carbon Vacancies of Graphene: A Trend Study from First Principles. *Phys. Rev. B,* **2013**, *87*(1), 014102.
[http://dx.doi.org/10.1103/PhysRevB.87.014102]

[22]   Xi, J.; Xie, C.; Zhang, Y.; Wang, L.; Xiao, J.; Duan, X.; Ren, J.; Xiao, F.; Wang, S. Pd Nanoparticles Decorated N-Doped Graphene Quantum Dots@N-Doped Carbon Hollow Nanospheres with High Electrochemical Sensing Performance in Cancer Detection. *ACS Appl. Mater. Interfaces,* **2016**, *8*(34), 22563-22573.
[http://dx.doi.org/10.1021/acsami.6b05561] [PMID: 27502735]

[23]   Rad, A.S.; Abedini, E. Chemisorption of NO on Pt-Decorated Graphene as Modified Nanostructure Media: A First Principles Study. *Appl. Surf. Sci.,* **2016**, *360*, 1041-1046.
[http://dx.doi.org/10.1016/j.apsusc.2015.11.126]

[24]   Kunaseth, M.; Mudchimo, T.; Namuangruk, S.; Kungwan, N.; Promarak, V.; Jungsuttiwong, S. A DFT Study of Arsine Adsorption on Palladium Doped Graphene: Effects of Palladium Cluster Size.

Appl. Surf. Sci., **2016**, *367*, 552-558.
[http://dx.doi.org/10.1016/j.apsusc.2016.01.139]

[25]   Jin, N.; Han, J.; Wang, H.; Zhu, X.; Ge, Q. A DFT Study of Oxygen Reduction Reaction Mechanism over O-Doped Graphene-Supported Pt4, Pt3Fe and Pt3V Alloy Catalysts. *Int. J. Hydrogen Energy,* **2015**, *40*(15), 5126-5134.
[http://dx.doi.org/10.1016/j.ijhydene.2015.02.101]

[26]   Gui, Y.; Peng, X.; Liu, K.; Ding, Z. Adsorption of C_2H_2, CH_4 and CO on Mn-doped graphene: Atomic, electronic, and gas-sensing properties. *Physica E,* **2020**, *113959*.
[http://dx.doi.org/10.1016/j.physe.2020.113959]

[27]   Wanno, B.; Tabtimsai, C. A DFT Investigation of CO Adsorption on VIIIB Transition Metal-Doped Graphene Sheets. *Superlattices Microstruct.,* **2014**, *67*, 110-117.
[http://dx.doi.org/10.1016/j.spmi.2013.12.025]

[28]   Zhou, Q.; Wang, C.; Fu, Z.; Tang, Y.; Zhang, H. Adsorption of Formaldehyde Molecule on Stone-Wales Defected Graphene Doped with Cr, Mn, and Co: A Theoretical Study. *Comput. Mater. Sci.,* **2014**, *83*, 398-402.
[http://dx.doi.org/10.1016/j.commatsci.2013.11.036]

[29]   Cortés-Arriagada, D.; Villegas-Escobar, N.; Ortega, D.E. Fe-Doped Graphene Nanosheet as an Adsorption Platform of Harmful Gas Molecules (CO, CO_2, SO_2 and H_2S), and the Co-Adsorption in O_2 Environments. *Appl. Surf. Sci.,* **2018**, *427*(2), 227-236.
[http://dx.doi.org/10.1016/j.apsusc.2017.08.216]

[30]   Cortés-Arriagada, D.; Villegas-Escobar, N.; Ortega, D.E. Fe-Doped Graphene Nanosheet as an Adsorption Platform of Harmful Gas Molecules (CO, CO_2, SO_2 and H_2S), and the Co-Adsorption in O_2 Environments. *Appl. Surf. Sci.,* **2018**, *427*, 227-236.
[http://dx.doi.org/10.1016/j.apsusc.2017.08.216]

[31]   Rao, D.; Wang, Y.; Meng, Z.; Yao, S.; Chen, X.; Shen, X.; Lu, R. ScienceDirect Theoretical Study of H_2 Adsorption on Metal-Doped Graphene Sheets with Nitrogen-Substituted Defects. *Int. J. Hydrogen Energy,* **2015**, *40*(41), 14154-14162.
[http://dx.doi.org/10.1016/j.ijhydene.2015.08.107]

[32]   Chen, X.; Chen, S.; Wang, J. Screening of Catalytic Oxygen Reduction Reaction Activity of Metal-Doped Graphene by Density Functional Theory. *Appl. Surf. Sci.,* **2016**, *379*, 291-295.
[http://dx.doi.org/10.1016/j.apsusc.2016.04.076]

[33]   Kunaseth, M.; Mudchimo, T.; Namuangruk, S.; Kungwan, N.; Promarak, V.; Jungsuttiwong, S. Ac Ce Pt e d Cr T. *Appl. Surf. Sci.,* **2016**.
[http://dx.doi.org/10.1016/j.apsusc.2016.01.139]

CHAPTER 8

Graphene Based Nanomaterials as Catalyst in Reduction Reactions

Leena Khanna[1,*], **Mansi** [1] and **Pankaj Khanna**[2]

[1] *University School of Basic and Applied Sciences, Guru Gobind Singh Indraprastha University, Sector 16-C, Dwarka, New Delhi-110078, India*

[2] *Department of Chemistry, Acharya Narendra Dev College, University of Delhi, Kalkaji, New Delhi-110019, India*

Abstract: The exceptionally outstanding physical and chemical properties as well as unique morphology of graphene have led to the development of various graphene-based catalysts, which are highly effective and selective in the reduction and hydrogenation reactions of organic compounds. This chapter is dedicated to compilation of the versatile reactions of hydrogenation/reduction over graphene-based catalysts. The use of catalyst allows highly effective and selective reduction of substrates in an effortless, recyclable, constructible and environmentally benign system.

Keywords: Eco-friendly, Graphene, Hydrogenation, Nanocomposites, Reduction, Solid support.

INTRODUCTION

The chemistry of graphene has recently been explored and become an important part of material science just after a breakthrough work done by Geim and Novoselov in 2004 [1 - 3]. It has a 2D-sheet structure having conjugated carbon atoms with sp^2-hybridization and an extended honeycomb-like network structure. Various properties of graphene, like high surface area, fine size, chemical inertness, great mechanical strength, and conductivity make it an ideal material for catalysis, organic conversion energy storage, *etc.*

The 2-dimensional single-layer carbon sheet structure of graphene serves as a building unit for the synthesis of graphite, fullerenes and nanotubes with three-, one- and zero-dimensional structures, respectively. Graphene sheets with a large

* **Corresponding author Leena Khanna:** University School of Basic and Applied Sciences, Guru Gobind Singh Indraprastha University, Sector 16-C, Dwarka, New Delhi-110078, India; E-mail: leenakhanna@ipu.ac.in

Manorama Singh, Vijai K. Rai and Ankita Rai (Eds)

surface area and of high quality can be achieved from silicon carbide (SiC) by their epitaxial single-crystal growth formation using ultrahigh vacuum annealing [4].

The developing concern regarding graphene is due to its environmentally benign nature and exceptionally wide properties, like high mechanical stiffness, unusual electrical conduction, elasticity and morphology [5 - 7]. However, it is easy to rapidly synthesize it. The extent of manufacturing of graphene derivatives, for *e.g.* graphene oxide, GO and reduced GO, RGO, provides extensive opportunities to process graphene-based multi-functional materials having use either as surface support for any catalyst or metal-free catalyst [8]. Graphene oxide (GO) is thoroughly examined due to its unique features, like remarkable electrical conductivity, elasticity, higher mechanical rigidity and morphology.

Graphene-based nanocomposites have wide-ranging applications in various fields, including drug delivery, sensors, energy, adsorption, electrochemistry, catalysis and environmental protection. There are many categories of carbon-based materials that have been used for catalytic applications, including graphene (GR), carbon nanotubes (CNT), graphitic carbon nitride (GCN), and activated carbon (AC) [8].

It is well known that the high catalytic activity of metal nanoparticles is because of their high surface area. However, practical applications of these nanoparticles are mainly faced with two main issues. The first issue is that they easily get agglomerated during a chemical reaction due to high surface energy, resulting in a decrease in their catalytic activity. The second issue is their small particle size, due to which their separation from the reaction is difficult. To overcome these disadvantages, the binding of these nanoparticles on a suitable solid surface support has been performed [9, 10]. Many solid surface supports have been employed for this binding, among which graphene and its derivatives are gaining importance [11, 12].Graphene-based nanocomposites are being used as a catalyst for the removal or treatment of organic pollutants present in wastewater [13]. Nowadays, in various organic transformations, there is a rapid increase in the use of graphene as an alternative to heterogeneous catalysts [14 - 17]. The aim of this chapter is to summarize the applicability of graphene-based nanomaterials, particularly in the reduction reactions of organic compounds.

HYDROGENATION OF C-C DOUBLE BONDS BY USING METAL-FREE GRAPHENE

It has been a challenging task to promote hydrogenation reactions with metal-free catalysts. Currently, the use of carbon materials for the hydrogenation of ethylene and nitroaromatics is gaining high importance [8]. A series of Gr materials, such as GO, Gr, rGO, and some doped Gr with heteroatoms, have been prepared and used for hydrogenation reactions. TEM images showed a high degree of hexagonal structural domains while atomic force microscopy (AFM) images depicted monolayer or few-layer morphologies of the Gr-based materials suspended in different solvents, like water, ethanol, *etc.* These Gr-based carbocatalysts were examined for selectively hydrogenating acetylene in the presence of excess ethene at room temperature to 150°C, in a stream containing thrice the excess amount of hydrogen as compared to acetylene (Scheme **1**). Nitrogen was applied as the carrier gas. The dopant elements used were not found to be not present at the active sites responsible for hydrogenation reaction. This was confirmed by comparing the reactivity rates of a series of non-doped Gr or rGO and heteroatom-doped Grs (it indicated that doped heteroatoms in Gr, such as N, O, S, and P, had little effect on the reactivity of Gr).

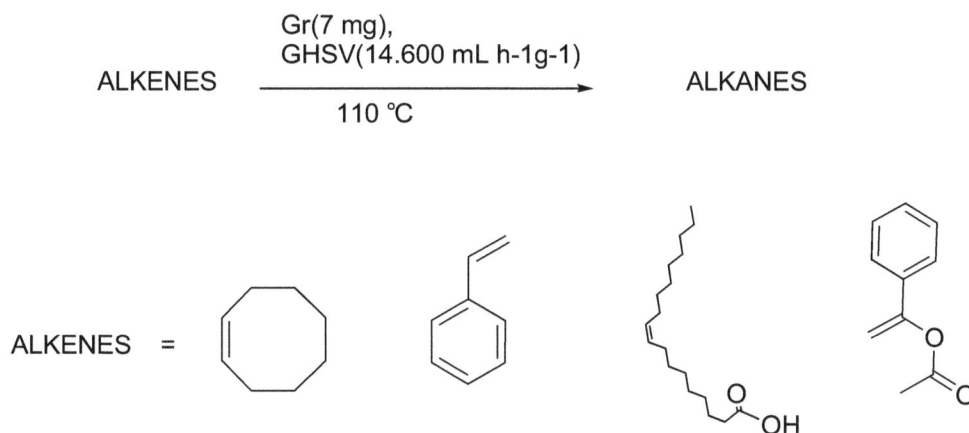

Scheme. (1). Hydrogenation of alkenes.

Reduction Using Ni-oxide/GOSs Catalyst

Nickel oxide/graphene composite, *i.e.* NiO/Graphene composite, was prepared by simple reducing or capping agent-free 'mix and heat' method. NiO nanoparticles were ultrafine spherical in shape, having an average size of 7.5nm [12]. They were uniformly dispersed on graphene oxide and firmly attached to GOSs surface, as revealed by TEM and AFM results. GO sheets were not aggregated and found in a single layer form. The elemental analysis of C, O, and Ni confirmed the homogenous dispersion of NiO NPs over the graphene oxide surface. This catalyst showed outstanding catalytic performance in the reduction of nitrophenol to amino phenols. A very low amount of catalyst (0.5 and 1 mg) is required for the reduction of nitrophenols at a high reaction rate and in a small reaction time (Scheme **2**). Also, these Ni-oxide/GOSs have shown much better efficacy in the reduction of carbonyl compounds using 2-propanol as an H donor. Reduction of benzaldehyde to benzyl alcohol was performed by the use of various bases, such as NaOH, KOH and K_2CO_3; the best results were provided by NaOH and it afforded excellent yield of the product (Scheme **3**).

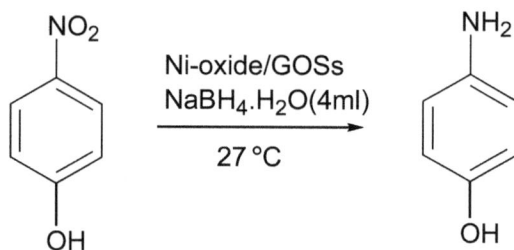

Scheme. (2). Reduction of nitrophenol.

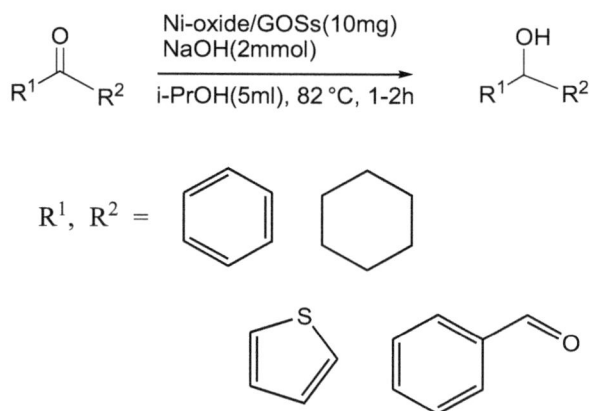

Scheme. (3). Reduction of carbonyl compounds.

Hydrogenation By Using Pd@CGO Catalyst

Nakhate *et al*. synthesized carbon-based graphene oxide monolith (Pd@CGO monolith), supporting palladium nanoparticles following solvothermal carbonization of glucose with a small amount of graphene oxide (GO) [18]. Broad peaks in the XRD pattern confirmed poorly arranged and crumpled sheets that were loosely packed together in CGO. Hence, monoliths were integrated by carbonaceous material from glucose, uniformly dispersed over the graphene oxide thin film sheet. It has a high flat surface area for attaching Pd NPs. The Pd NPs were irregular in shape, having particle size 4-20 nm but homogeneously dispersed over the CGO surface. The irregular shape of Pd NPs displayed that the catalyst was recyclable with the least decrease in catalytic efficiency. Pd@CGO monolith catalyst has been utilized further in the reduction of nitrobenzene (Scheme **4a**) and hydrogenation of alkenes (Scheme **4b**) by using H_2. Various substitutions on nitrobenzenes and alkenes could also be hydrogenated. This process showed a high conversion rate towards the required product. Compounds like n-dodecene, cyclohexene, cinnamaldehyde and 2-nitrotoluene were efficiently hydrogenated by using H_2 at 1 atm pressure, n-decane as an internal standard and IPA as solvent (20 ml) at 50°C. The catalyst has been reused four times.

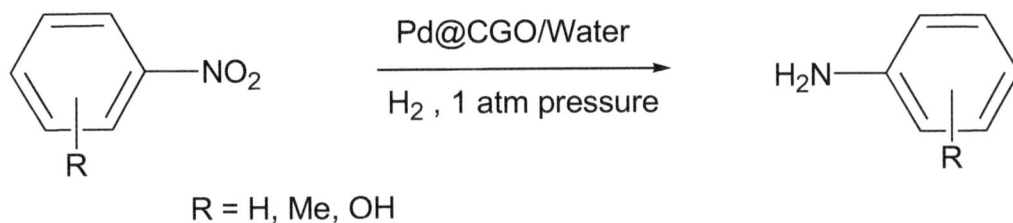

Scheme. (4a). Reduction of Nitrobenzene.

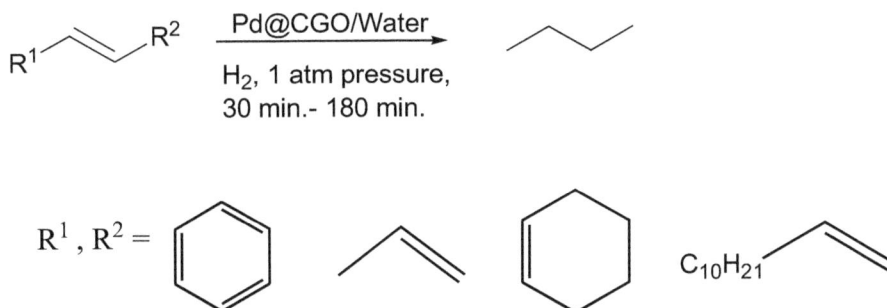

Scheme. (4b). Hydrogenation of alkenes.

Reduction Of Nitroarenes Into Aniline Derivatives By Au-GO Catalyst

Choi *et al.* synthesized Au-GO hybrid NPs and evaluated their reduction capability in nitroarenes using reductant sodium borohydride ($NaBH_4$) under ambient reaction conditions (Scheme **5**) [19, 20]. The frequency turnover ranged from 300-2400/s, describing the effectively concentrated active AuNPs on the surface of graphene sheets [20, 21]. The single-layer graphene nanosheets formed a colloidal suspension of GO, having a thickness of about 0.70 nm and lateral dimensions of 0.70-1.5 mm, as confirmed by atomic force microscopy (AFM). The synergistic effect of AuNPs and GO showed by the Au-GO nanostructure and their catalytic reduction activity were highly stable. Zhu *et al.* synthesized GSCN/Au NPs hybrid nanocatalyst; at first, graphene oxide/SiO_2 composite nanosheets (GSCN) were prepared through the sol-gel process, and then on their surface, small well-dispersed Au NPs were assembled [21]. GSCN showed good dispersity having sheet-like morphology with an average thickness of ca. 23.3 ± 0.6 nm. SiO_2 acted as a stabilizer on the surface of graphene, preventing the aggregation of GO occurring due to strong van der Waals interaction between the GO sheets. The GSCN acted as a supporting material for AuNPs due to the high specific surface area, stability and solubility, thereby facilitating the reduction of nitrobenzene. In another work, Gao *et al.* [22] have used a simple RGO catalyst for the room temperature reduction of nitrobenzene. The high surface area of graphene nanosheets, used as a catalytic support for AuNPs, accelerated the reaction because of their distinctive electronic structural edges. The catalytic activity of NPs elevated by the synergistic effect of support and helped in the reduction of nitroarenes, thereby providing high yields.

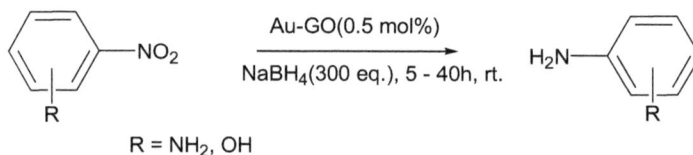

$$R = NH_2, OH$$

Scheme. (5). Reduction of nitroarenes.

Reduction By Ru-GCN Catalyst

Sharma *et al.* established an excellent heterogeneous photocatalyst (Ru-GCN) and used it efficiently for the transfer of hydrogen to nitroarenes (Scheme **6**), carbonyls (Scheme **7**) and alkenes (Scheme **8**), and *via* alcohols/hydrazine mediation under 9W LED domestic lamp [23, 24]. It has a high turnover number (TON) and provided very good yields of alcohol, alkanes and amines, respectively. Under visible light irradiation, it has been reported to be a sustainable and simple method, and is used for the reduction of carbonyls. GCN is the most stable allotrope of all carbon nitrides at ambient conditions, contains basic surface functionalities and is able to undergo many chemical reactions due to its electron-rich nanostructures. GCN (g-C_3N_4) of large surface area with high pore volume can be derived from urea. Further, Ru-g-C_3N_4 has been prepared with an agglomeration of these small units of g-C_3N_4 with dense morphology. It has been screened for selective transfer hydrogenation reaction involving nitroarenes and carbonyls. The reaction exhibited excellent catalytic activity in visible light, using hydrazine hydrate as a hydrogen donor under mild reaction conditions. The catalyst was easy to separate and was found to be non-toxic, environmentally friendly and with a high turn-over frequency. This catalyst limits the use of external energy sources and high-pressure hydrogenation reactors.

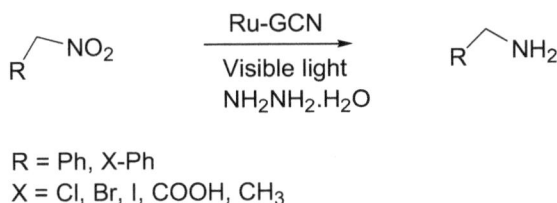

R = Ph, X-Ph
X = Cl, Br, I, COOH, CH₃

Scheme. (6). Reduction of nitro compounds.

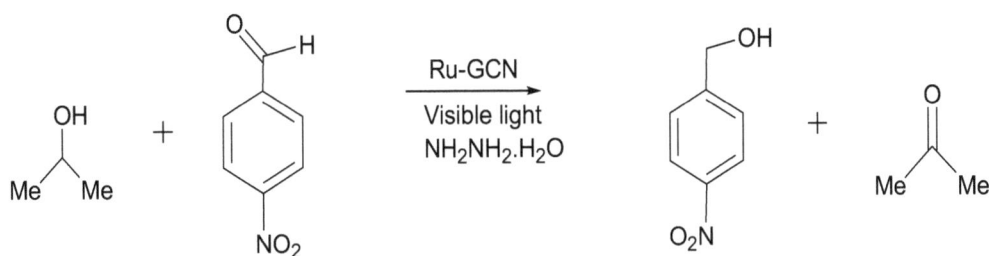

Scheme. (7). Reduction of carbonyl compounds.

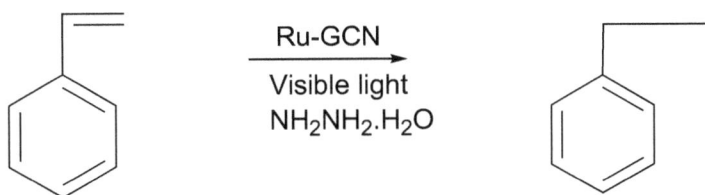

Scheme. (8). Hydrogenation of alkenes.

Hydrogenation By Using G-Pt Catalyst

Sheng *et al.* prepared a G-Pt catalyst in which Pt NPs having an average diameter of 2-3 nm were vastly dispersed on graphene sheets [25]. In this methodology, Pt ions were reduced on the GO surface to form small and fine Pt NPs with no significant agglomeration. The catalyst was used in the hydrogenation of nitroarene to the corresponding amines (Scheme 9) as well as for hydrogenation of alkenes and alkynes (Scheme 10), showing excellent catalytic activity. The presence of electron-withdrawing/electron-donating groups did not influence the nitroarene reduction. Also, the reactions completed in 1h. Simple platinum nanoparticles (Pt NPs) often show low catalytic activity after prolonged use due to agglomeration in the reaction system. However, these G-Pt composites have been found to be more superior as a catalyst in the hydrogenation of various unsaturated compounds at 1 bar initial pressure of hydrogen and under mild reaction conditions with CH₃OH as the solvent.

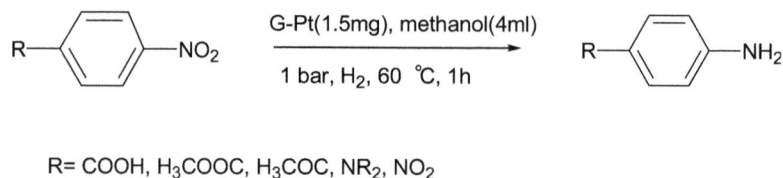

R= COOH, H₃COOC, H₃COC, NR₂, NO₂

Scheme. (9). Reduction of nitro compounds.

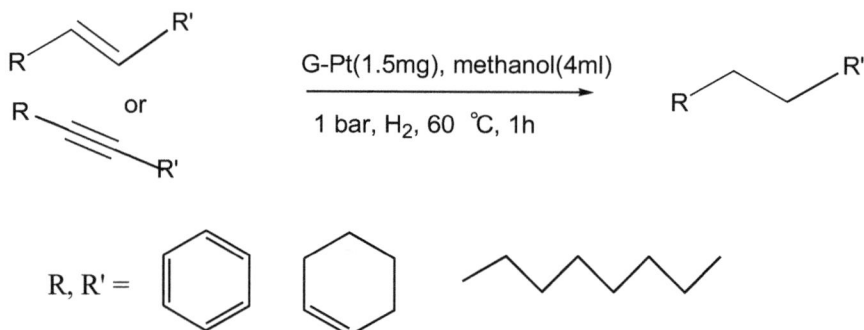

Scheme. (10). Hydrogenation of alkenes and alkynes.

Reduction By Using NrGO Catalyst

Ahmad *et al.* reported the synthesis and utility of NrGOs [26] (Scheme **11**). The morphological studies revealed a layered structure with two to three layers of graphene sheets. GO sizes ranged from 1.0-1.5 nm in thickness, and BET surface area was 481 m^2/g. The selective hydrogenation of the nitro moiety (yield 65%) in the presence of alkenes or alkynes has been achieved using the NrGO catalyst (Scheme **12a** and **12b**) [27]. Nitroarenes in the presence of NrGO, base t-BuOK and solvent IPA, have been selectively reduced by H_2 gas at 1.5 MPa, when passed into a glass tube containing them, equipped with a steel-based autoclave. The reaction completed in 2h on heating at 130°C. The addition of t-BuOK in NrGO helped in the neutralization of acidic functionality and quickened the localization of radicals at basic sites. The scope of NrGO catalyst has also been extended to the reduction of carbonyl compounds (Scheme **13a** and **13b**). The chemoselectivity for nitro hydrogenation is ~68% over the carbonyl group. The radical species promoted the electron transfer to substrates and generated hydrogen radicals. The ionic moiety of NrGO selectively interacted with more polar nitro groups as compared to less polar groups, like alkenes, alkynes, carbonyls, *etc*. The strong interaction of nitro groups with these NrGOs promoted their selective reduction.

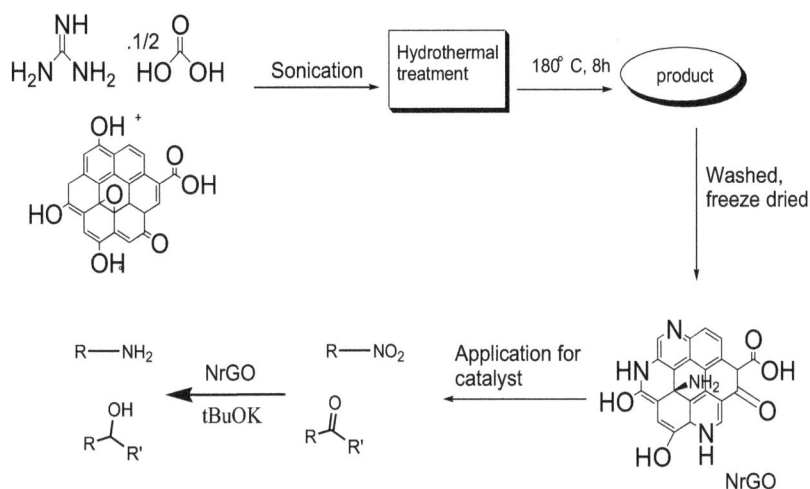

Scheme. (11). Schematic illustration of the synthesis of NrGO and its application in catalysis.

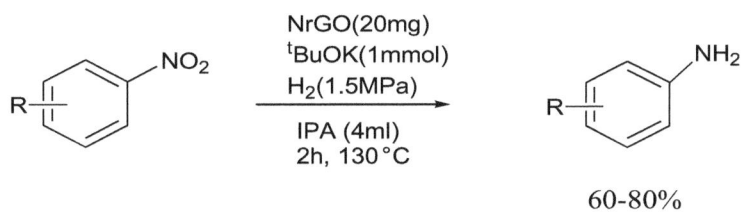

60-80%

Here,

R = Br, I, Cl, CF₃, OMe, Me, NH₂, OH

Scheme. (12a). Reduction of nitroarenes.

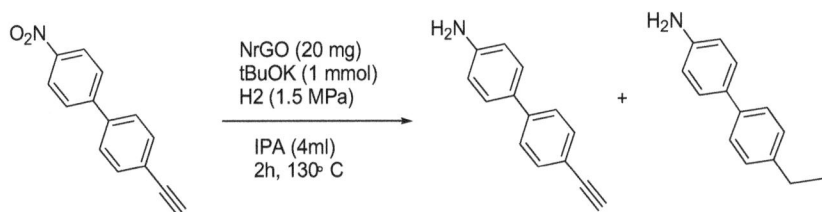

Scheme. (12b). Chemoselective hydrogenation of nitroarenes.

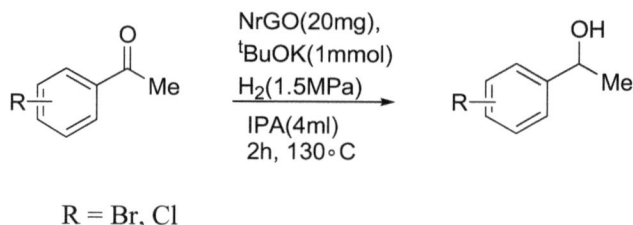

R = Br, Cl

Scheme. (13a). Reduction of carbonyls.

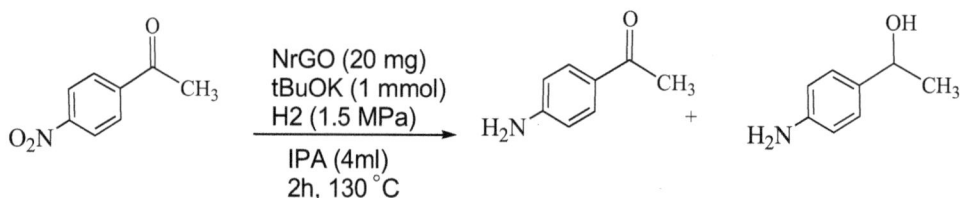

Scheme. (13b). Chemoselective hydrogenation of carbonyl and nitroarenes.

Reduction Of 4-Nitrophenol Into Aniline Derivatives By Using AuNP/PQ11/GN Catalyst

AuNPs were integrated into composite materials (GNs) to explore their various applications. PQ11, a cationic polyelectrolyte, was used as a polymeric reducing agent of $HAuCl_4$ and a stabilizing agent for GNs to form AuNPs decorated over GNs [28]. The average thickness of a GO sheet was measured to be 0.863 nm. However, the average thickness of a single PQ11/GN was measured to be 2.964 nm. It was found that several nanoparticles were attached to the surface of GN. The size of the AuNPs ranged from 10 to 80 nm, and the shape was irregular. These AuNP/PQ11/GN nanocomposites displayed fine catalytic activity in the reduction of 4-nitrophenol (Scheme **14**).

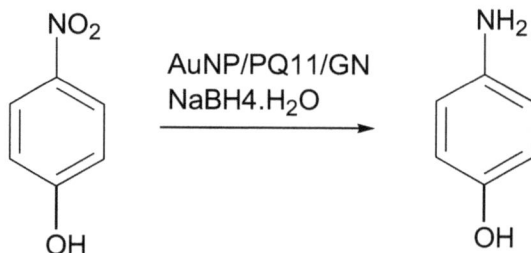

Scheme. (14). Reduction of nitrophenols.

Hydrogenation Of Carbonyl Compounds By Using Ru/RGO Catalyst

Hydrogenation of carbonyl compounds at low temperature is a highly challenging task; however, it is a significant process for producing fuels and chemicals. The major problem is the lack of heterogeneous catalysts that can work at low temperatures. In this cadre, reduced graphene oxide supported Ru nanoparticles (Ru/RGO) showed remarkable ability towards low-temperature hydrogenation of carbonyl group in carbonyl compounds into C-OH group [29]. Ru/RGO involves a BET surface area of $324.09 m^2/g$ and diameter of 2.0 nm, and is uniformly dispersed in nature. Many compounds, like 2-pentanone, hydroxyacetone, furfural, propionaldehyde, cyclohexanone, acetone, acetophenone and benzophenone, have been reduced successfully using this catalyst (Scheme **15**). It has superior activity and selectivity at low-temperature (even at -10°C) in the hydrogenation of LA (levulinic acid). GVL was obtained at 20 °C using the catalyst in 99.9% yield with TOF of $2112.2 \ h^{-1}$. The excellent catalytic activity was credited to the easy transfer of an electron between Ru (0) to RGO, leading to the generation of electron-rich Ru (0) nanoparticles, which were highly reactive to activate the C=O group towards reduction.

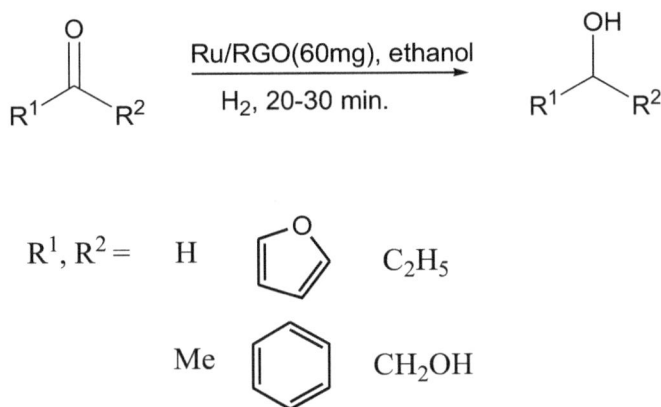

Scheme. (15). Reduction of carbonyl compounds.

Fe NPs/CDG Mediated Hydrogenation Of 1-Hexene

The iron catalyst has been rarely examined for the hydrogenation of alkenes. Synthesis of iron nanoparticles supported graphene (CDG) was performed by decomposition of Fe $(CO)_5$ using ultrasonic waves, andFe-NP/CDG was obtained [30]. This catalyst contained Fe NPs with an average size of 4.37 ± 1.62 nm deposited on the CDG sheets, and its application as a hydrogenation catalyst was studied. It was found that hydrogenation reaction on CDG in the presence of EtMgCl or $MgCl_2$, without Fe-NPs, provided nil hydrogenated products. However, Fe-NP/CDG catalyst when used provided the hydrogenated product of terminal and cyclic olefins in very high yields (Scheme **16**). The catalyst could be easily recycled several times and separated by magnetic decantation. It was proposed that GR reduced remaining oxide shells of Fe NPs acting as a surface activator, which were formed on CDG *via* remaining oxide functions.

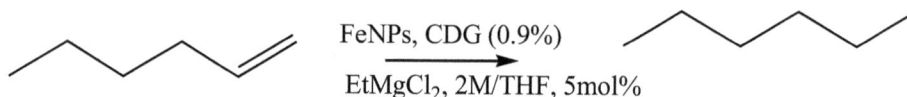

Scheme. (16). Hydrogenation of alkenes.

Hydrogenation Of Styrene By Using Fe@g-C₃N₄ Catalyst

It has been found that the formation of desired products using metal-free GCN is highly difficult as it is required to overcome the energy barrier. The nanocomposites of GCN are used a sustainable tool for heterogeneous catalysis with magnetic nanoparticles [31]. The immobilization of iron ferrite by using GCN was firstly examined by Baig *et al.* In this case, immobilized iron oxide over the cavities of GCN was used to obtain the Fe@GCN catalyst. For organic transformations, a synergic effect has been exerted by Fe-oxide NPs on the GCN surface. The synthesized Fe@GCN catalyst with hydrazine hydrate was used for the hydrogenation of alkenes and alkynes under mild conditions without any requirement of hydrogenation at high pressure (Scheme **17**).

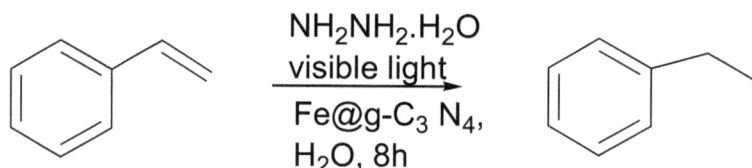

Scheme. (17). Hydrogenation of styrene.

Hydrogenation Of Phenylacetylene By Using Pd@mpg-C₃N₄ Catalyst

Another heterogeneous catalyst consisting of mesoporous GCN supporting Pd NPs (Pd@mpg-C₃N₄) was developed by Wang *et al.* for the liquid-phase partial hydrogenation of phenylacetylene under mild reaction conditions (303 K, atmospheric H₂) [32]. Pd/mpg-C₃N₄ possesses a surface area of 120 m²/g, pore volume of 0.28cc/g and a mesopore diameter of 38.1 nm. This catalyst was also found to be effective in the hydrogenation of many alkynes as well with good recyclability. A chemoselective product (styrene) was obtained in 94% yield within 85 min (Scheme **18**). The solid support materials played an important role in the action of the Pd metal for the present reaction. The mpg-C₃N₄ provided large specific surface areas and electron density. This helped in the interaction of palladium with the support, distribution and morphology of the Pd ions and desorption of the product formed after the reaction.

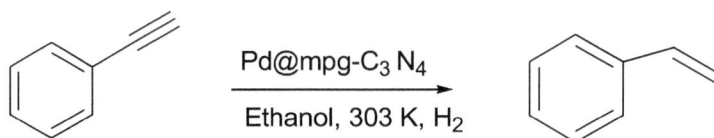

Scheme. (18). Hydrogenation of phenylacetylene.

Hydrogenation Of Alkynes And Terminal Alkynes By Using Co@NGR Catalyst

Co@NGR catalyst was prepared having cobalt nanoparticles of size 10-50 nm. They were spherical in shape, very well supported and distributed over the graphene sheets [33]. Co_3O_4 phase was in well accordance and the lattice spacing in Co@NRG was found to be 0.204 nm, which confirmed the presence of Co (0) species in Co@NRG. The selected-area electron diffraction (SAED) image showed a ring-like diffraction pattern which indicated the particles to be partially crystalline. The highly stable N-doped Co@NGR nanocatalyst was used for the hydrogenation of alkynes, more specifically for Z-alkenes (Scheme **19**). The substituents, like flouro, methyl, methoxy, chloro, *etc.*, on terminal alkynes could be hydrogenated smoothly. The reaction proceeded well *via* the transfer of hydrogen and provided the corresponding vinyl alkenes in 91-99% yield under optimal conditions. The catalytic activity was found to be outstanding even after nine runs. The proposed mechanism has been described below (Scheme **20**).

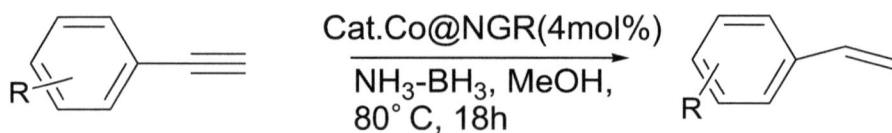

R = F, Cl, Me, ^{t}Bu, OMe

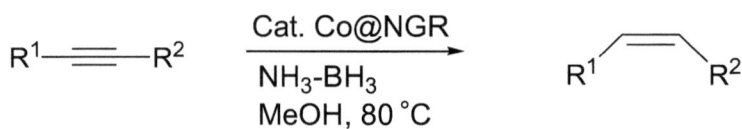

symmetrical bisaryl alkenes:

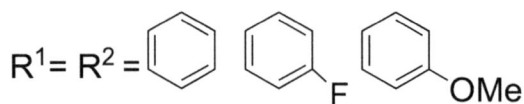

unsymmetrical alkenes:

R^1 = aryl
R^2 = Alkyl

R^1= R^2 =

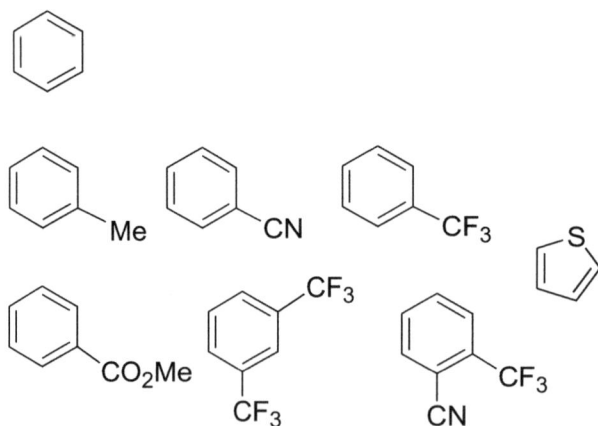

Scheme. (19). Hydrogenation of alkynes.

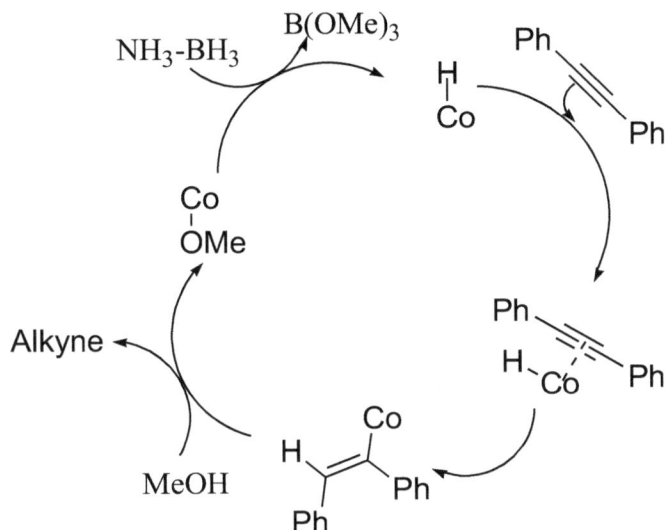

Scheme. (20). Schematic of the catalytic cycle.

HYDROGENATION OF ALKYNES BY USING SGR/PANI/NI CATALYST

Polyaniline/graphene oxide nanocomposites were prepared by liquid–liquid interface polymerization reaction of aniline using iron (III) chloride and hydrogen peroxide as oxidants in the suspension of sulfonated graphene oxide [34]. Further, NiNPs were decorated on this synthesized hybrid material. The stability of SGR/PANI/Ni was more in comparison to SGR/PANI and degraded slowly only at high temperatures due to the presence of thermally stable nickel metal. The SGR/PANI was amorphous and well-dispersed over the sulfonated graphene oxide. The agglomerated features on the SGR/PANI composite indicated the homogeneous dispersion of NiNPs. SAED pattern of SGR/PANI/Ni showed various differences in the crystal planes, indicating that NiNPs were completely ingrained inside the SGR/PANI composite. This resultant synthesized hybrid catalyst SGR/PANI/Ni showed excellent catalytic activity towards selective hydrogenation of phenylacetylene with its derivatives (Scheme **21**). The reaction was performed under mild temperature and hydrogen pressure at 100 psi was applied. The catalyst was successfully recycled with a minimum decrease in catalytic activity. Its application was also studied for the hydrogenation of terminal alkynes to selective alkenes.

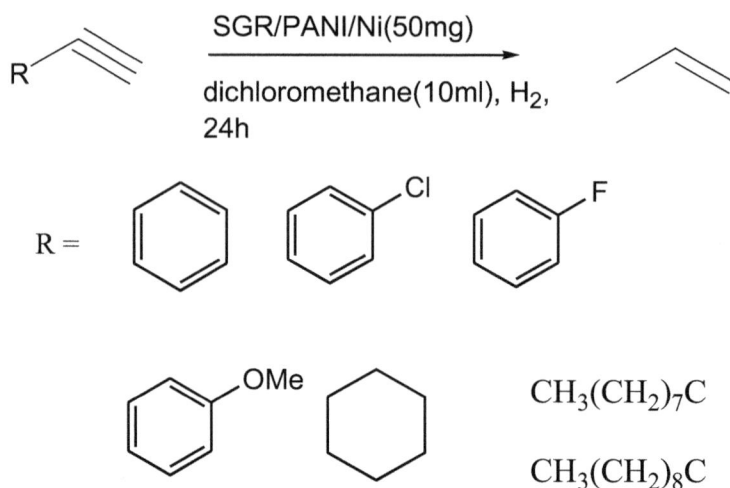

Scheme. (21). Hydrogenation of terminal alkynes.

Hydrogenation Of Alkenes And Alkynes By Using rGO-Ni$_{30}$Pd$_{70}$

The bimetallic Pd alloy NPs, Ni$_{30}$Pd$_{70}$, supported by reduced graphene oxide showed high catalytic performance in comparison to their monometallic counterparts. There was a synergistic effect observed out of two different metal interactions [35]. The colloidal Ni$_{30}$Pd$_{70}$ NPs were deposited on rGO through a liquid phase self-assembly method. TEM images of colloidal Ni$_{30}$Pd$_{70}$ NPs and rGO-Ni$_{30}$Pd$_{70}$ catalysts displayed uniform particle distribution of the alloy NPs with 3.5 nm of average particle size. The exclusive role of rGO was to enhance the activity of these alloy NPs, which was done either by well-dispersion of NPs over the rGO nanosheets or the presence of a graphitic plane (2D) around each rGO-NiPd catalyst. This helped to pull organic reactants in the near-range contact with the NPs. The catalytic performance of these reduced graphene oxide supported Ni$_{30}$Pd$_{70}$ alloy NPs (rGO-Ni$_{30}$Pd$_{70}$) was studied for the direct hydrogenation of unsaturated hydrocarbons to alkanes (Scheme **22**). This catalyst surpassed Pd/C commercial catalyst both in terms of stability and activity. A variety of cyclic or aromatic compounds, alkenes and alkynes, were speedily reduced to the corresponding saturated products in high yields (>99%) under ambient conditions.

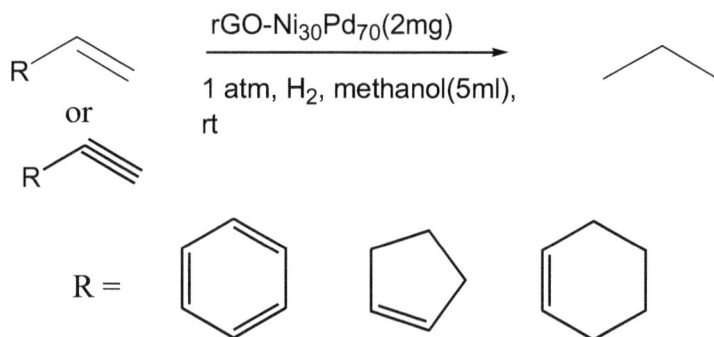

Scheme. (22). Hydrogenation of alkenes and alkynes.

Hydrogenation Of Double Bonds By Using Pd-NGRO-$_{300}$ Catalyst

N-doped reduced graphene oxides (NRGO) were obtained by following urea-hydrothermal treatment of graphene oxide, supporting Pd nanoparticles. The N of NRGO improved the diffusion of Pd-NPs. The Pd/NRGO$_{-300}$ catalyst formed was extremely reactive and selectively hydrogenated the double bond of cinnamaldehyde and also hydrogenated the ion of phenol to cyclohexanone under moderate reaction conditions [36] (Scheme **23**). The catalyst has been found to be highly recyclable for many runs without any decrease in reactivity. NRGO was intended to fasten very fine Pd NPs (~1.6 nm), homogeneously dispersed over the thin graphene layers, with increased interaction between Pd and N in NRGO. Pd NPs involved a lattice space of 0.224 nm and a well-crystallized (111) face. The homogeneous nature of N doping over the entire graphene sheets was studied by elemental maps and line scans of C and N.

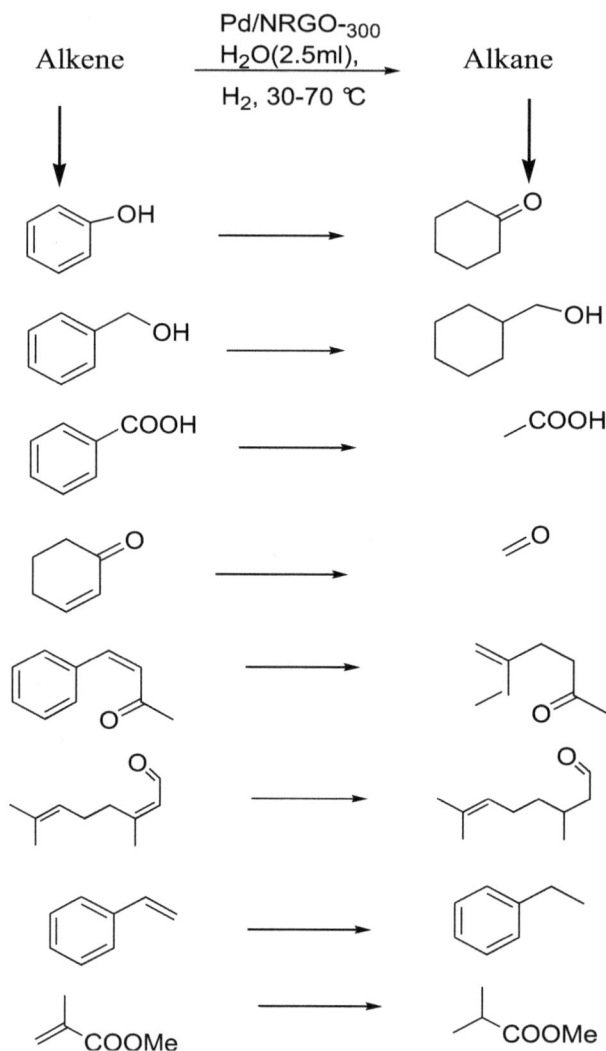

Scheme. (23). Hydrogenation of alkenes.

CONCLUSION

This chapter has briefly reviewed the applications of graphene-based nanocatalysts in hydrogenation reactions. The graphene-based nanocatalysts may or may not contain metals. In particular, the focus has been on the hydrogenation of nitro, carbonyl and alkenes or alkynes and synthesis of amines, alcohols and alkanes, respectively, in excellent yield. Many different types of carbon-based nanocomposites have been explained, such as graphene (GR), carbon nanotubes (CNT), graphitic carbon nitride (GCN), and activated carbon (AC), which are used for organic reductions as well as other applications involving coupling reactions and oxidation reactions.

Distinctive description of each type of carbon-based nanocomposites has been summarized with a specific example, having some morphological features such as pore size, surface area, and shape of nanoparticles with chemical selectivity. It has been observed that due to some extraordinary features like large surface area, high mechanical strength, along with high conductivity and excellent catalytic activity in hydrogenation reactions, these nanocatalysts are playing a role of stepping stone for future organic chemistry.

CONSENT FOR PUBLICATION

Not applicable.

CONFLICT OF INTEREST

The author declares no conflict of interest, financial or otherwise.

ACKNOWLEDGEMENTS

The authors are thankful to Guru Gobind Singh Indraprastha University, New Delhi, for funding under Faculty Research Grant Scheme (FRGS).

REFERENCES

[1]   Novoselov, K.S.; Geim, A.K.; Morozov, S.V.; Jiang, D.; Zhang, Y.; Dubonos, S.V.; Grigorieva, I.V.; Firsov, A.A. Electric field effect in atomically thin carbon films. *Science,* **2004**, *306*(5696), 666-669.
       [http://dx.doi.org/10.1126/science.1102896] [PMID: 15499015]

[2]   Chen, Y.; Zhang, B.; Liu, G.; Kang, E.; Chen, Y. Chem Soc Rev Graphene and its derivatives : switching on and off. *Chem. Soc. Rev.,* **2012**, *41*(13), 4585-4772.
       [http://dx.doi.org/10.1039/c2cs35043b]

[3]   Wan, X.; Huang, Y.; Chen, Y. Focusing on energy and optoelectronic applications: a journey for graphene and graphene oxide at large scale. *Acc. Chem. Res.,* **2012**, *45*(4), 598-607.
       [http://dx.doi.org/10.1021/ar200229q] [PMID: 22280410]

[4]   Berger, C.; Song, Z.; Li, X.; Wu, X.; Brown, N.; Naud, C.; Mayou, D.; Li, T.; Hass, J.; Marchenkov, A.N.; Conrad, E.H.; First, P.N.; de Heer, W.A. Electronic confinement and coherence in patterned epitaxial graphene. *Science,* **2006**, *312*(5777), 1191-1196.
       [http://dx.doi.org/10.1126/science.1125925] [PMID: 16614173]

[5]   Raccichini, R.; Varzi, A.; Passerini, S.; Scrosati, B. The role of graphene for electrochemical energy storage. *Nat. Mater.,* **2015**, *14*(3), 271-279.
       [http://dx.doi.org/10.1038/nmat4170] [PMID: 25532074]

[6]   Kumar, P.; Shahzad, F.; Yu, S.; Man, S.; Kim, Y.; Min, C. Large-area reduced graphene oxide thin film with excellent thermal conductivity and electromagnetic interference shielding effectiveness. *Carbon N. Y.,* **2015**, *94*, 494-500.
       [http://dx.doi.org/10.1016/j.carbon.2015.07.032]

[7]   Paredes, J.I. S. V.-R. Biomolecule-assisted exfoliation and dispersion of graphene and other two-

dimensional materials:a review of recent progress and applications. *R. Soc. Chem.,* **2016**, 1-68.

[8]   Primo, A.; Neatu, F.; Florea, M.; Parvulescu, V.; Garcia, H. Graphenes in the absence of metals as carbocatalysts for selective acetylene hydrogenation and alkene hydrogenation. *Nat. Commun.,* **2014**, *5*, 5291.
        [http://dx.doi.org/10.1038/ncomms6291] [PMID: 25342228]

[9]   Tian, Z.; Li, Q.; Hou, J.; Li, Y. S. A. Highly selective hydrogenation of α, β-unsaturated aldehydes by Pt catalysts supported on Fe-based layered double hydroxides and derived mixed metal oxides. *Catal. Sci. Technol.,* **2016**, *6*, 703-707.
        [http://dx.doi.org/10.1039/C5CY01864A]

[10]   Wang, W.; Gu, J.; Hua, W.; Jia, X.; Xi, K. A novel high efficiency composite catalyst: single crystal triangular Au nanoplates supported by functional reduced graphene oxide. *Chem. Commun. (Camb.),* **2014**, *50*(64), 8889-8891.
        [http://dx.doi.org/10.1039/C4CC03306J] [PMID: 24968811]

[11]   Bej, A.; Ghosh, K.; Sarkar, A. D. W. K. Palladium nanoparticles in the catalysis of coupling reactions. *RSC Advances,* **2013**, 1-15.

[12]   Saravanamoorthy, S.; Vijayakumar, E.; Jemimahc, S. A. I. Catalytic Reduction of p-nitrophenol and carbonyl compounds by NiO-nanoparticles fastened graphene oxide. *Chem. Sci. Eng. Res.,* **2019**, *1*, 1-7.
        [http://dx.doi.org/10.36686/Ariviyal.CSER.2019.01.01.001]

[13]   Hu, M.; Yao, Z.; Wang, X. Graphene-based nanomaterials for catalysis. *Ind. Eng. Chem. Res.,* **2017**, *56*, 3477-3502.
        [http://dx.doi.org/10.1021/acs.iecr.6b05048]

[14]   Hadi Abedi, M.M. Synthesis of three-metal layered double hydroxide and dual doped graphene oxide composite as a novel electrocatalyst for oxygen reduction reaction - ScienceDirect. *J. Power Sources,* **2016**, *333*, 53-60.
        [http://dx.doi.org/10.1016/j.jpowsour.2016.09.152]

[15]   Pendashteh, A.; Palma, J.; Anderson, M.; Marcilla, R. Applied catalysis B : environmental NiCoMnO$_4$ nanoparticles on n-doped graphene : highly efficient bifunctional electrocatalyst for oxygen reduction / evolution reactions. *Appl. Catal. B,* **2017**, *201*, 241-252.
        [http://dx.doi.org/10.1016/j.apcatb.2016.08.044]

[16]   Low, J.; Yu, J.; Ho, W. Graphene-based photocatalyst for CO reduction to solar fuel submit it to jpc letter as a perspective paper graphene-based photocatalyst for CO$_2$ reduction to solar fuel. *J. Phys. Chem. Lett.,* **2015**, 1-26.

[17]   Cheng, Y.; Lin, J.; Xu, K.; Wang, H.; Yao, X.; Pei, Y.; Yan, S.; Qiao, M.; Zong, B. Fischer − tropsch synthesis to lower ole Fe Ns over potassium- promoted reduced graphene oxide supported iron catalysts. *ACS Appl. Mater. Interfaces,* **2016**, *6*, 389-399.

[18]   Nakhate, A.V.; Yadav, G.D. Palladium nanoparticles supported carbon based graphene oxide monolith as catalyst for sonogashira coupling and hydrogenation of nitrobenzene and alkenes. *ChemistrySelect,* **2016**, *2*(1), 3954-3965.
        [http://dx.doi.org/10.1002/slct.201600819]

[19]   Choi, Y.; Bae, S.; Seo, E.; Jang, S.; Park, H.; Kim, B. Hybrid gold nanoparticle-reduced graphene oxide nanosheets as active catalysts for highly efficient reduction of nitroarenes. *J. Mater. Chem.,* **2011**, *21*, 15431-15436.
        [http://dx.doi.org/10.1039/c1jm12477c]

[20]   Garg, B.; Ling, Y.; Geim, S. Versatilities of graphene-based catalysts in organic transformations. *green Mater.,* **2012**, *1 (GMATI)*, 47-61.

[21]   Zhu, C.; Han, L.; Hu, P. S. D. Nanoscale graphene oxide / SiO$_2$ composite nanosheets and their catalytic properties. *Nanoscale,* **2012**, *4*, 1641-1646.

[http://dx.doi.org/10.1039/c2nr11625a] [PMID: 22286065]

[22]   Gao, Y.; Ma, D.; Wang, C.; Guan, J.; Bao, X. Reduced graphene oxide as a catalyst for hydrogenation of nitrobenzene at room temperature. *Chem. Commun. (Camb.),* **2011**, *47*(8), 2432-2434.
[http://dx.doi.org/10.1039/C0CC04420B] [PMID: 21170437]

[23]   Bahuguna, A.; Kumar, A.; Krishnan, V. Carbon-support-based heterogeneous nanocatalysts : synthesis and applications in organic reactions. *Asian J. Org. Chem.,* **2019**, *8*, 1-44.
[http://dx.doi.org/10.1002/ajoc.201900259]

[24]   Sharma, P.; Sasson, Y. Highly active Ru-g-C$_3$N$_4$photocatalyst for visible light assisted selective hydrogen transfer reaction using hydrazine at room temperature. *Catal. Commun.,* **2017**, *102*(August), 48-52.
[http://dx.doi.org/10.1016/j.catcom.2017.08.019]

[25]   Sheng, B.; Hu, L.; Yu, T.; Gu, H. RSC Advances highly-dispersed ultrafine Pt nanoparticles on graphene as effective hydrogenation catalysts. *RSC Advances,* **2012**, *1*(C), 5520-5523.
[http://dx.doi.org/10.1039/c2ra20400b]

[26]   Ahmad, M.S.; He, H.; Nishina, Y. Selective hydrogenation by carbocatalyst : the role of radicals. *Org. Lett.,* **2019**, *21*(20), 8164-8168.
[http://dx.doi.org/10.1021/acs.orglett.9b02432] [PMID: 31584281]

[27]   Gao, Y.; Tang, P.; Zhou, H.; Zhang, W.; Yang, H.; Yan, N.; Hu, G.; Mei, D.; Wang, J.; Ma, D. Graphene oxide hot paper graphene oxide catalyzed C À H bond activation : the importance of oxygen functional groups for biaryl construction. *Angew. Chem. Int. Ed.,* **2016**, *99352*(55), 3124-3128.
[http://dx.doi.org/10.1002/anie.201510081]

[28]   Qin, X.; Lu, W.; Luo, Y.; Chang, G.; Asiri, A.M.; Al-Youbi, A.O.; Sun, X. Anchoring gold nanoparticles on graphene nanosheets functionalized with cationic polyelectrolyte: a novel catalyst for 4-nitrophenol reduction. *J. Nanosci. Nanotechnol.,* **2012**, *12*(4), 2983-2989.
[http://dx.doi.org/10.1166/jnn.2012.5818] [PMID: 22849055]

[29]   Tan, J.; Cui, J.; Cui, X.; Deng, T.; Li, X.; Zhu, Y.; Li, Y. Graphene-modified Ru nanocatalyst for low-temperature hydrogenation of carbonyl groups. *ACS Catal.,* **2015**, *5*, 7379-7384.
[http://dx.doi.org/10.1021/acscatal.5b02170]

[30]   Stein, M.; Wieland, J.; Steurer, P.; Tçlle, F.; Mülhaupt, R.; Breit, B. Iron nanoparticles supported on chemically-derived graphene : catalytic hydrogenation with magnetic catalyst separation. *Adv. Synth. Catal.,* **2011**, *353*, 523-527.
[http://dx.doi.org/10.1002/adsc.201000877]

[31]   Baig, R.B.N.; Verma, S.; Varma, R.S.; Nadagouda, M.N. Magnetic Fe @ G-C$_3$N$_4$:a photoactive catalyst for the hydrogenation of alkenes and alkynes. *ACS Sustain. Chem.& Eng.,* **2016**, *4*(3), 1661-1664.
[http://dx.doi.org/10.1021/acssuschemeng.5b01610]

[32]   Deng, Dongshun; Yangb, Yang; Gongb, Yutong; , YiLib; , Xuan Xub; , Y. W. Palladium nanoparticles supported on Mpg-C3N4 as active catalyst for semihydrogenation of phenylacetylene under mild conditions. **2013**, (207890), 1-6.

[33]   Jaiswal, G.; Landge, V.G.; Subaramanian, M.; Kadam, G.; Zbo, R.; Gawande, M.B.; Balaraman, E. N-graphitic modified cobalt nanoparticles supported n-graphitic modified cobalt nanoparticles supported on graphene for tandem dehydrogenation of ammonia- borane and semihydrogenation of alkynes. *ACS Sustain. Chem.& Eng.,* **2020**, 1-39.
[http://dx.doi.org/10.1021/acssuschemeng.9b07211]

[34]   Panwar, V.; Kumar, A.; Singh, R.; Gupta, P.; Ray, S.S.; Jain, S.L. Nickel-decorated graphene oxide / polyaniline hybrid : a robust and highly efficient heterogeneous catalyst for hydrogenation of terminal alkynes. *Ind. Eng. Chem. Res.,* **2015**, *54*, 11493-11499.
[http://dx.doi.org/10.1021/acs.iecr.5b02888]

[35]   Cetinkaya, Y.; Balci, M.; Metinc, O. Reduced graphene oxide supported nickel-palladium alloy nanoparticles as a superior catalyst for the hydrogenation of alkenes and alkynes under ambient conditions. *RSC Advances,* **2016**, *6*, 28538-28542.
[http://dx.doi.org/10.1039/C5RA25376D]

[36]   Nie, R.; Miao, M.; Du, W.; Shi, J.; Liu, Y.; Hou, Z. Selective hydrogenation of CC bond over N-doped reduced graphene oxides supported Pd catalyst. *Appl. Catal. B,* **2015**, 1-25.

Graphene Based Nanomaterials as Catalysts in Solar Water Splitting

Bilal Chikh[1], Imane Ghiat[1], Adel Saadi[1] and Amel Boudjemaa[1,2,*]

[1] Laboratory of Natural Gas, Faculty of Chemistry, USTHB, Algiers

[2] Centre de Recherche Scientifique et Technique en Analyses Physico-Chimiques, Bou-Ismail CP 42004, Tipaza, Algeria

Abstract: Graphene, a famous material, is rapidly rising in the field of materials science. The 2D material demonstrates in a small period of history particular properties, which are briefly discussed in the present chapter. This work reviews the preparation, characterization and potential applications of graphene in water splitting. The chapter starts with a short introduction to graphene, its properties, and structural features. Then, a section on the different methods used for its preparation and its modification are discussed in detail. In addition, this chapter contains some practical examples of graphene-based materials largely used for hydrogen generation. In addition, this chapter contains a thorough revision of the latest results on the use of graphene-based materials for hydrogen production *via* water splitting using photocatalytic or photoelectrocatalytic processes.

Keywords: 2D materials, Graphene, Graphene-based nanomaterials, Graphene oxide, Hydrogen, Nanomaterials, Photoelectrochemical, Reduced graphene, Water splitting.

INTRODUCTION

Graphene (Gr), the hyped 2D matrix of carbon atoms arranged in a hexagonal lattice, with a carbon-carbon distance of 0.142 nm [1] (See Fig. **1**). Gris the lightest, strongest, thinnest material known to man, as well as the best heat and electricity conductor ever discovered with intrinsic electron mobility up to 200,000 cm^2/Vs and high surface areas (~ 2600 m^2/g), *etc* [2, 3]. In addition, Gr is 200 times stronger than steel manufactured and it is one of the strongest materials ever tested [4]. Due to its incredible properties, Gr is used in many applications, prominently in the electronic, environment, medical, photonic, drug delivery, energy storage *etc.* [5 - 9].

* **Corresponding author Amel Baudjemaa:** Centre de Recherche Scientifique et Technique en Analyses Physico-Chimiques, Bou-Ismail CP 42004, Tipaza, Algeria, Algiers; Email: amel_boudjemaa@yahoo.fr

Manorama Singh, Vijai K. Rai and Ankita Rai (Eds)

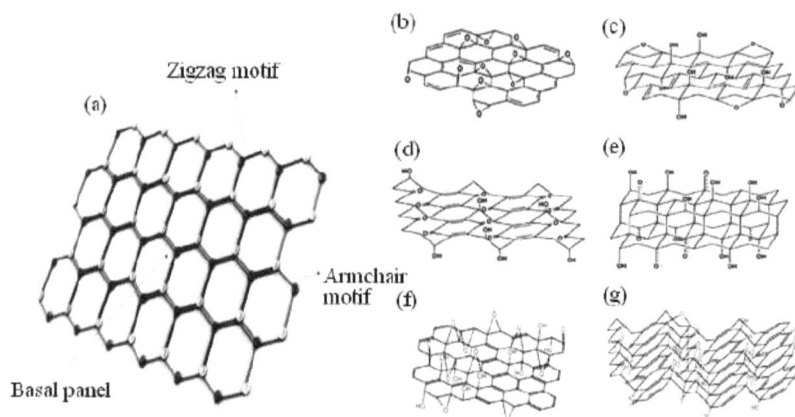

Fig. (1). Structures of pristine graphene (a) and its graphene oxide derivatives based on Hofmann (b), Ruess (c), Scholz-Boehm (d), Nakajima-Matsuo (e), Lerf-Klinowski (f) and Szabo (g) models [1, 9].

With a 2D planar single sheet of sp^2 hybridized carbon, Gr is a basic building block for other allotropes of carbon such as fullerene (0D), carbon nanotube (CNT) (1D), and graphite (3D) [10] (Fig. **2**). Gr represent a whole class of 2D materials, including single layers of Molybdenum-disulphide (MoS_2) and Boron-Nitride (BN) [11]. Gr is the subject of relentless research and it is the idea to be able to change the whole of industries. Gr has attracted huge attention since its first discovery by isolation from bulk graphite using adhesive tape [12 - 17]. Gr has exhibited remarquable properties as electronic, mechanical, thermal, optical, and optoelectronic properties such as ultrahigh carrier mobility, thermal conductivity, and mechanical strength [14 - 20]. Also, it can be wrapped up into fullerenes (0D), rolled into CNT (1D), or stacked into graphite (3D).

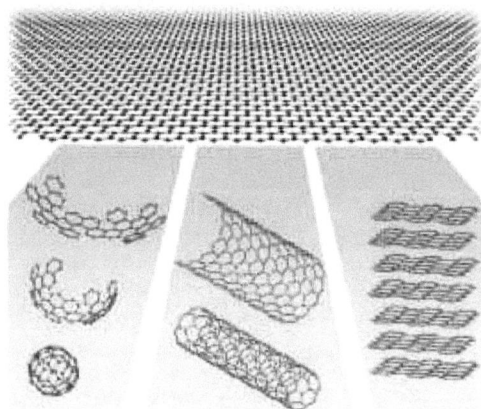

Fig. (2). Schematic demonstration of Gr as the mother of different kinds of carbon-based materials [10].

Since 2004, Gr has drawn much attention due to its remarkable properties [21 - 24]. It is reported that the theoretical value of the specific surface area of Gr reaches 2000 m^2/g [25] make it ideal support and improving their enhanced adsorption activity. It is also known that Gr has a robust but flexible structure with high carrier mobility. Therefore, Gr-based photocatalyst shows excellent photocatalytic properties. Moreover, it can be easily obtained Gr from graphite, which is cheap and naturally sufficient.

GRAPHENE: THE WONDER MATERIAL PROPERTIES AND SYNTHESIS

Graphene Properties

Graphite consists of layered sp^2 bonded carbon with strong in-plane bonds and weak bonding between the planes. Single-layer graphite, called Gr, is a novel material discovered in 2004 [11]. It can be viewed as the basic building block of all carbon allotropes (Figs. 2 and 3) and presents huge potential.

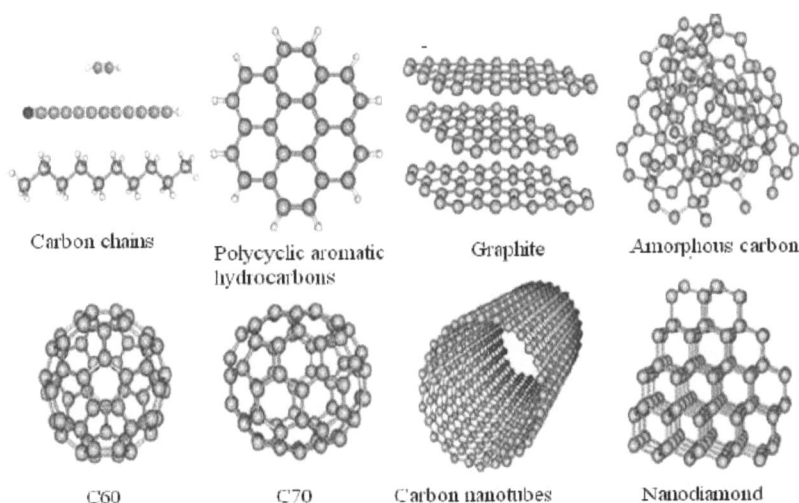

Carbon chains

Polycyclic aromatic hydrocarbons

Graphite

Amorphous carbon

C60

C70

Carbon nanotubes

Nanodiamond

Fig. (3). The diversity of carbon allotropes [26].

Applications [26]. The zero-energy bandgap [10] material commonly synthesized *via* mechanical cleavage of highly orientated pyrolytic graphite (HOPG) or chemical vapor deposition (CVD) can be treated within the theory of mass less Dirac fermions [27]. The analysis is closely related to that of topological insulators [28], where surface properties are different from the bulk behaviour. On the other hand, Gr has low spin-orbit coupling leading to large phase coherence such that a supercurrent has been found to flow through this two-dimensional (2D) material [29].

SYNTHESIS METHODS OF GRAPHENE

The number of Gr-related research works and patents has augmented in haste over the last five years, intimating the era of Gr prosperity, moreover, the statistics are expected to increase in the forthcoming years [30, 31]. Despite the fast expansions in Gr synthesis methods, a cost-effective method for mass production of high-quality Gr remains scarce and expresses a major challenge [32, 33]. Various methods have been created and new manufacturing processes are still emerging.

At present, the Gr synthesis methods can be classified into two groups, namely Top-Down (destruction) and Bottom-Up (construction) method [34, 35]. The Top-Down route involves the structural breakdown of graphite and other carbon-based precursors followed by the interlayer partition forming Gr sheets. The methods are mechanical exfoliation [36], arc discharge [37], oxidative exfoliation-reduction [38], liquid-phase exfoliation (LPE) [39] and unzipping of CNTs [40] comprise the Top-Down route. Meanwhile, the bottom-up technique comprised those methods, which use a carbonaceous gas source to produce Gr. The most relevant ones are chemical vapor deposition (CVD) [41, 42], template route [43], substrate-free gas-phase synthesis (SFGP) [44], epitaxial growth [45] and total organic synthesis [46].

TECHNIQUES OF CHARACTERIZATION FOR GRAPHENE

Gr has emerged as the most popular topic in the active research field since Gr's discovery in 2004 by Andrei Geim and Kostya Novoselov. Since then, Gr research has exponentially accelerated because of its extraordinary properties, which have attracted the interest of researchers all over the world. Therefore, a number of reviews have documented the outstanding properties such as morphological, textural, and surface chemistry of Gr that were achieved by several characterization techniques (Fig. **4**). the present part will discuss the major characterization techniques of Gr, including Raman spectroscopy, transmission electron microscopy (TEM), scanning electron microscopy (SEM), atomic force microscopy (AFM), and others.

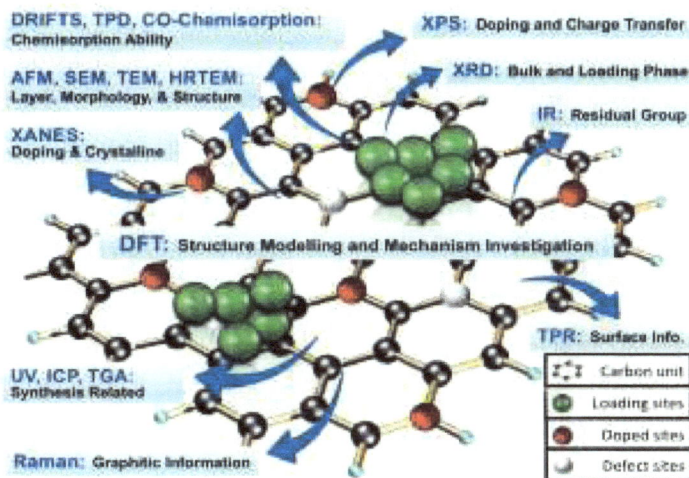

Fig. (4). Schematic diagram of characterization techniques for graphene-based materials [47].

Fourier Transform Infrared Spectroscopy (FTIR)

FTIR spectroscopy is a versatile technique for the characterization of materials belonging to the carbon family. Based on the interaction of the IR radiation with the matter this technique may be used for the determination of chemical functional groups in Gr and its derivatives. The most important features of this method are non-destructive, real-time measurement and relatively easy to use. Typically, graphite has characteristic peaks at 3430 and 1610 cm^{-1} corresponding to O–H stretching of adsorbed water and skeletal vibration from graphitic domains of aromatic C = C, respectively (Fig. **5**) [48, 49].

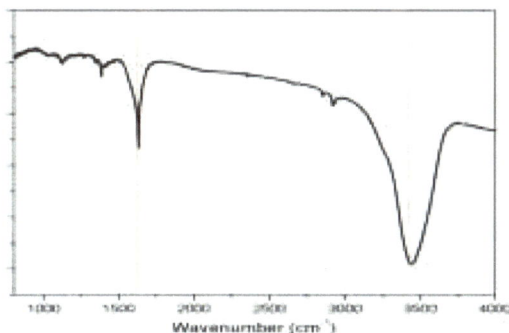

Fig. (5). FTIR spectrum of Gr sheets [49].

Raman Spectroscopy

Raman spectroscopy is viewed as one of the most important characterization tools in Gr research and it has been used as a non-destructive technique to characterize

Gr film. Carbon allotropes possess unique Raman characteristic peaks known as D-, G- and 2D-peaks at approximately 1350, 1580 and 2700 cm^{-1} [50]. In general, the G-peak represents the tangential stretching (E_{2g}) mode of highly oriented pyrolytic graphite while the D-peak refers to the disorder in sp^2 hybridized carbon atoms (lattice distortion) and the 2D-peak is the second-order Raman scattering process. The intensity ratio of D- peak to G-peak (I_D/I_G) can be used to determine the degree of disorder in the Gr sample. Meanwhile, the quality of Gr can be evaluated by calculating the ratio of the intensity of 2D/G (I_{2D}/I_G) from the Raman spectrum. A large ratio of I_{2D}/I_G and comparably minor amplitude of D peak implies that good-quality Gr has been produced [51]. The group of Li reported that the production of Gr by CVD led to a higher I_{2D}/I_G as compared to that elaborated by mechanical exfoliation [52]. Therefore, CVD is the preferable choice to synthesize high-quality Gr.

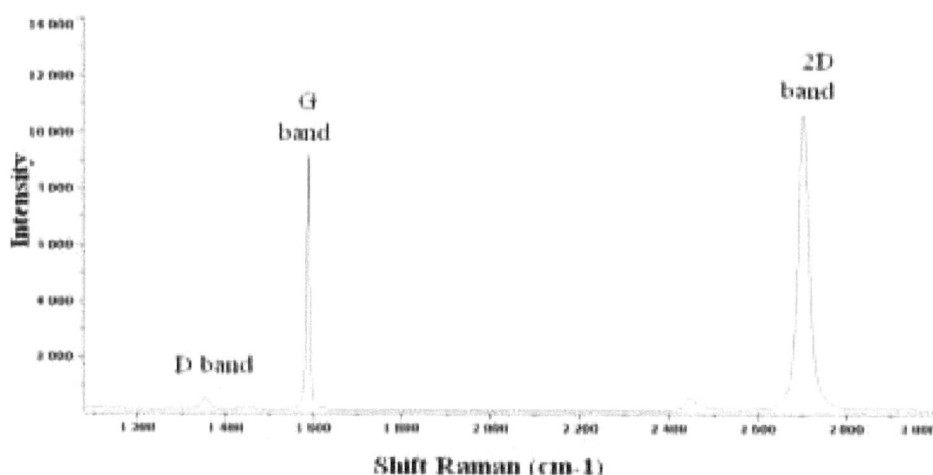

Fig. (6). Raman spectrum (excited by 514.532 nm laser) of exfoliated Gr sheets [53].

Atomic Force Microscopy (AFM)

Due to its convenience and reliability, AFM can be used to measure the thickness, surface roughness and other morphological features of Gr and its derivatives [54]. In a typical measurement, Gr was scanned by the mean of a cantilever with a sharp tip (~5–10 nm). The subtle changes in the heterogeneity of the surface allow the collection of the sensitive changes in vibration amplitude and frequency of the tip, which can analyze the topography of the sample. The obtained three-dimensional images thus enable the measurement of the thickness of Gr films. In consideration of that, the perfect Gr has a single atom thickness of around 0.35 nm, thus, the number of the layers could be counted. For instance, the group of You measured the thickness and the surface roughness of Gr film *via* AFM for gas sensor applications to determine the best growth condition. The results revealed that for long CVD reaction times, the surface roughness and the thickness of the Gr increased, which was explained by the formation of more layers of Gr [55]. The obtained AFM images are shown in Fig. (**7**). Unfortunately, this technique provides only topographical images has difficulty in scanning bulky material [56].

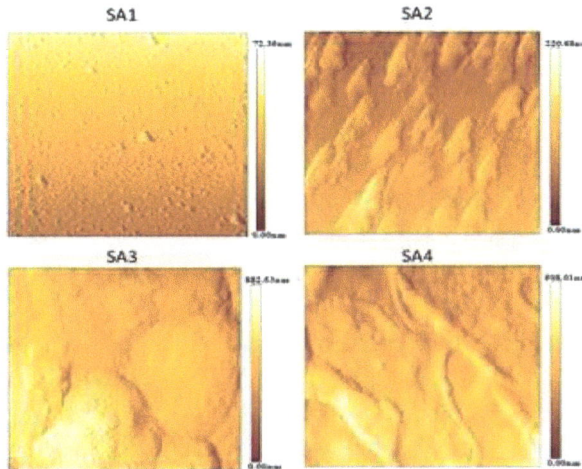

Fig. (7). The micrographs of SA1, SA2, SA3 and SA4 obtained in the close contact mode. Size: 3μm×3μm [55].

Scanning Electron Microscopy (SEM)

SEM is a technique used for examining the morphology of Gr and its derivatives. The Gr material is placed on a silicon wafer of specific thickness for SEM scanning, and the number of layers can be estimated based on the color depth [57]. For instance, the group of Tu used SEM to analyze Gr film prepared *via* CVD process on a Cu substrate [58]. The obtained SEM images show continuous Gr with a few white areas of wrinkles on the surface, as observed in Fig. (**8**).

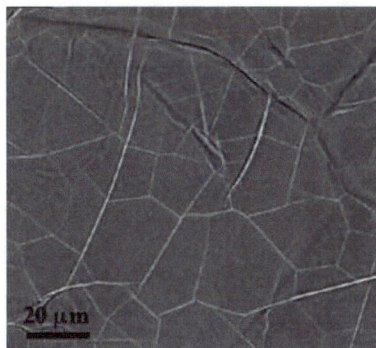

Fig. (8). SEM image of Gr sheet of 7 layers on a copper substrate [58].

Transmission Electron Microscope (TEM)

TEM is one of the most versatile methods to analyze Gr due to its ability to image the material down to the atomic scale. However, the TEM characterization is very important in Gr research since it can be used to identify point defects, Stone–Wales rotation, vacancy, dislocations and many more [59, 60]. In addition, the thickness, number of layers and cross-sectional view of folded Gr sheets at different locations could be determined by the mean of high-resolution TEM (HRTEM). In TEM, a beam of energetic electrons is transmitted through a sample and the interaction of electrons with the sample forms an image. Monolayer Gr can be regarded as a transparent sheet using TEM characterization. Meanwhile, multi-layered Gr sheets are shown as multiple dark lines. Besides, the typical hexagonal crystalline nature of Gr can by identified by selected area electron diffraction (SAED). Fig. (9) illustrates the low- and high-magnification TEM images of a monolayer Gr film transferred onto the TEM grid, along with its corresponding SAED pattern extracted from the work of Tu *et al.* [60]. The TEM micrographs (Fig.9 (a)-(f)) clearly show2–7 Gr layers, while the SAED patterns in Fig. 9 (g)–(i) were found to be irregular. Based on these patterns, the justification of the bilayer Gr, trilayer Gr and five-layer Grfilms won't be possible. Thus, other characterizations, such as Raman spectroscopy, are crucial to support the TEM results.

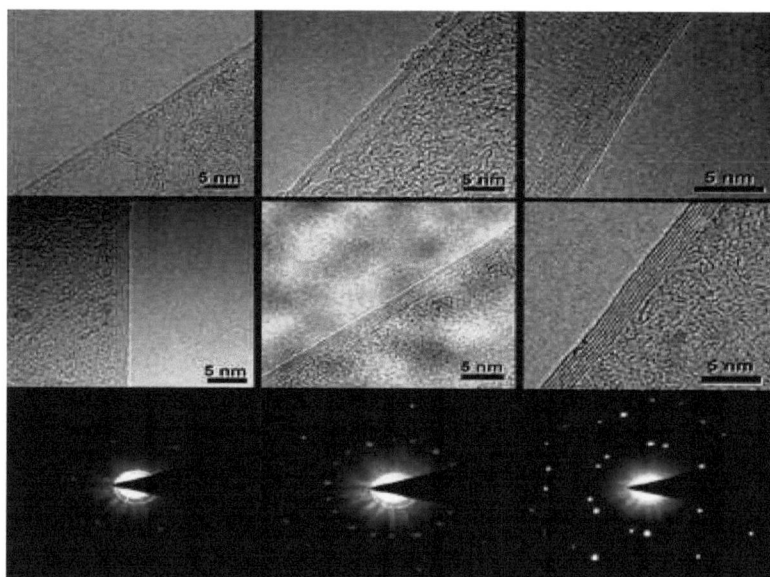

Fig. (9). High-resolution TEM images of the edges of Gr with different numbers of layers:(a) bilayer,(b) trilayer, (c) four layers, (d) five layers, (e) six layers and(f) seven layers.The typical SAED images of bilayer, trilayer and five-layer Gr taken from the centre of the domains are shownin(g)–(i), respectively [58].

GRAPHENE FOR WATER SPLITTING APPLICATION

Owing to the high surface of the area of Gr, it is an excellent support material for semiconducting photocatalyst materials. Furthermore, the defective cites present in Gr-based materials act as a nucleation center for the semiconducting metal oxide nanoparticles. Similarly, the Gr-based material, specifically reduced Gr oxide (rGO) with a negatively charged surface, exhibits an attractive material for the metal oxides. Due to this attractive force, the metal oxide nanoparticle becomes immobile, which prevents the agglomeration of the metal oxide semiconducting nanoparticles that improve the durability of the photocatalyst and also enhance the material reusability.

Water Splitting Process

Photocatalytic water splitting is a clean and renewable process that used natural sources; sunlight and water. The conversion of the water to hydrogen using a photocatalytic process has been described as an ideal solution to answer the environmental issues associated with fossil fuels [61]. This process was inspired by natural photosynthesis in green plants and is known as the Z-scheme.

H_2 production *via* water splitting was reported for the first time by Fujishima and Honda using a photo-electrochemical cell with TiO_2 photoanode [62]. TiO_2 has been largely used as a photocatalyst because it is stable, abundant, low cost, non-corrosive and environmentally friendly material. More importantly, its energy levels are appropriate to initiate the water-splitting reaction [63], where the conduction band (CB) of TiO_2 is more negative than the reduction energy level of water ($E_{H+/H2} = 0$ V), while the valance band (VB) is more positive than the oxidation energy level of water ($E_{O2/H2O} = +1.23$ V) (Fig. 10). The water-splitting equation is as follows [64]:

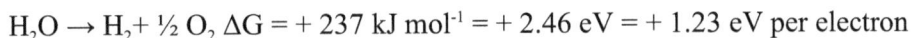

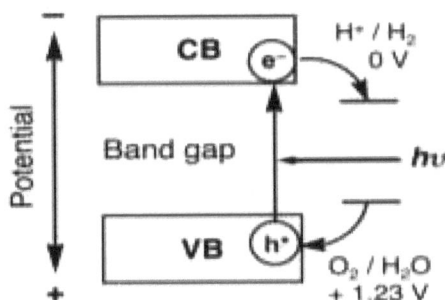

$$H_2O \rightarrow H_2 + \tfrac{1}{2}\, O_2 \quad \Delta G = +237 \text{ kJ mol}^{-1} = +2.46 \text{ eV} = +1.23 \text{ eV per electron}$$

Fig. (10). Mechanism of the photocatalytic water splitting [65].

For this process, the photocatalyst should have a suitable thermodynamic potential for water splitting reaction, suitably narrow bandgap energy (Eg) to harvest the visible photons and stability against photo corrosion [65].

Photocatalytic Water Splitting

The photocatalytic water splitting under visible light was demonstrated for the first time in 2001 using a Z-scheme photocatalytic system where $SrTiO_3$ doped Cr/Ta was used for H_2 evolution and WO_3 for O_2 generation, and an iodate/iodide (IO_3^-/I^-) redox couple as an electron mediator (Fig. **11**) [66]. In the case of photocatalytic water splitting, the photocatalyst is used as powders. The photocatalysts are dispersed in the aqueous solutions under irradiation. When, the material is irradiated with light (h$v \geq$ Eg) greater or equal to the band gap energy of the material, the electron in the VB can be excited to the CB leaving a hole (h^+) in the VB. Thus, the formation of the photogenerated (e^-/h^+) pairs is occurred. When an electron falls from the CB to holes in the VB, the recombination occurs [67]. The electrons photogenerated can migrate to the photocatalyst surface and reduce water to hydrogen gas, while the h^+ acts as an oxidizing agent to oxidize water to oxygen [24, 34]. In the addition of the photocatalyst reactivity, the reaction efficiency depend on the (i) pH of the solution, (ii) the photocatalyst amount, (iii) the particle size of the photocatalyst and (iv) the light intensity. The formation of electrons in the CB and positive holes in the VB depend on excitation by photon, which has equal or higher energy than the bandgap of semiconductor material [68, 69]. In photocatalysis, recombination of the pairs (e^-/h^+) is the major problem. Since photogenerated e^- and h^+ in their exciting form are unstable, excited e^- can be rapidly recombined with h^+ on the VB. Consequently, the energy is lost as heat.

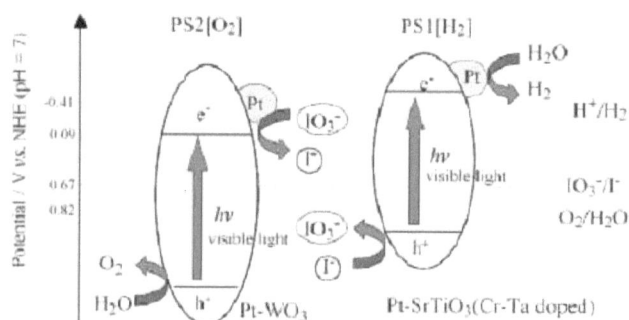

Fig. (11). Speculated reaction mechanism for the water splitting using a mixture of Pt-SrTiO$_3$, Pt-WO$_3$ and NaI aqueous solution [66].

Photoelectrochemical (PEC) Water Splitting

The PEC cell composed of working electrode (WE), Pt auxiliary electrode and a saturated calomel electrode (SCE). Generally, the W. E. was prepared using the elaborated photocatalyst, which was prepared as pellets or past [70, 71]. Fig. (12) illustrates the photoelectrochemical (PEC) water splitting based on n-type semiconductor [72]. The pairs (e^-/h^+) are generated when the photoanode absorbs photons. The e^- photoexcited transfer to a counter electrode and reduce water to H_2, while the h^+ transfer to the photoanode surface and oxidize water to O_2. So, the development of stable and efficient photoelectrode for PEC water splitting under solar light has been widely investigated. Usually, oxides photocatalysts as photoanodes combined with counter electrodes have been used for H_2 generation due to the high stability against photocorrosion [71, 73]. Generally, the CB levels of metal oxides are more positive for H_2 evolution. So, it is desirable to develop a photocatalyst that has an amply high band level for H_2 generation and a narrow bandgap energy (Eg < 3.0 eV) for visible light absorption.

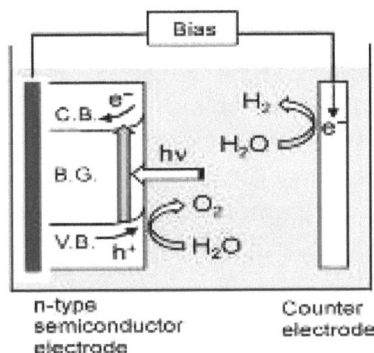

Fig. (12). Photoelectrochemical water splitting systems using n-type semiconductor photoanode [72].

Gr and Gr-based nanomaterials are considered the most promising photocatalyst due to their zero-band gap [68]. The Eg band of the semiconductors can be tuned with Gr [74]. In addition, Gr can restrain the (e^-/h^+) recombination *via* its strong electron mobility, where the photoreactivity can be improved. In the photocatalysis, the 2D structure of Gr gives a face to face orientation between photocatalyst is an important property [12].

On the other hand, used Gr with a photoactive photocatalyst can extend the light absorption in a large region of solar energy [75]. Thus, GO is more important than pristine Gr in the Gr-based photocatalyst due to its low-cost production. In addition, more than 2000 publications of Gr-based photocatalyst materials have been reported.

For several decades H_2 generation *via* by photocatalytic process has been followed over a variety of materials with a variety of structures and compositions, but only a few have given evidence that a catalytic reaction indeed occurs based on the suppression of recombination of the pairs (e^-/h^+) that is one of the major problems in the field of photocatalysis. For this, the choice of a suitable photocatalyst is an important step for efficient hydrogen evolution. On the other hand, after the first publication of Fujishima and Honda, many researchers have studied this subject gained great attention in science. But, due to the absorption of TiO_2 in the UV region, the research focuses on the synthesis of a new nanosized photocatalyst capable of absorption under visible light [76 - 81].

Graphene Based Nanomaterials

TiO$_2$-Graphene Photocatalysts

Gr as a new carbon nanomaterial, has unique properties, such as a large surface area, a good electron mobility, high transparency, and a flexible structure [14, 10, 15 - 18, 82]. Thus, the combination of TiO_2 and Gr is promising to enhance the photocatalytic properties of TiO_2. TiO_2 is largely used for hydrogen (H_2) generation *via* photocatalytic water splitting process due to its low cost, non-toxicity, and high stability [83, 84]. On the other hand, TiO_2-Gr demonstrated an effective photocatalytic activity for the protection of the environment as organic compounds elimination, air treatment and COV removal [85 - 89]. So, coupling of TiO_2 and Gr presents many advantages in terms of photocatalytic activity: (i) Gr can trap and move charge carriers, thus avoiding recombination of the pairs (e^-/h^+); (ii) the composite can be easily recovered from aqueous solution due to the large Gr sheets; (iii) the gap energy tuning and/or extension of the excitation wavelength can be achieved [90].

Thus, combined TiO_2 with carbon materials such as fullerenes (C60), CNT and Gr has been considered as a significant way to enhancement the photocatalytic activity of TiO_2 [91], especially for energy conversion applications. The excellent properties of Gr indicated that it has enormous potential to be an ideal construction component of photocatalytic materials for H_2 generation [92]. Compared to other applications of Gr, a small number of reports have been reported the application of Gr in the area of H_2 evolution. The Gr is used like an electron acceptor and the transporter has been classified by various photocatalytic tests.

Zhang and co-workers investigated the H_2 generation over TiO_2-Gr composite prepared by a sol-gel method using Na_2S and Na_2SO_3 as sacrificial agents [93]. TiO_2-Gr showed higher photocatalytic than P25. It has been reported that graphene oxide (GO) can be reduced to reduced graphene (rG) by solvothermal reaction of GO in the ethanol solvent [94]. Numerous studies established that the electron photo-generated migrate from the conducting band of TiO_2 semiconductor to Gr, which can efficiently suppress the recombination of photoinduced carriers and promote photocatalytic H_2 evolution [95]. In addition, the extremely high conductivity of Gr can make the accepted electrons migrate rapidly across its 2D plane to active sites [96]. Thus, Gr served as an acceptor of the photogenerated electrons of TiO_2 and transporter to separate the photogenerated pairs (e^-/h^+), in fact, the increase of the lifetime of the charge carriers.

A comparison with a commercial TiO_2 (P25) was investigated by Zhang and co-authors for the photocatalytic H_2 generation under UV irradiation [97]. So, the H_2 amount increase from 4.5 to 5.4 μmol h^{-1} when the Gr content increased from 0.8 to 2.0 wt%, respectively (Fig. **13**). This result is due to the high electron conductivity of Gr. So, the unique features of Gr allow photocatalytic reactions to take place not only on the surface of semiconductor catalysts but also on the Gr sheet. But, the increase in Gr amount to 5 wt% shows a decrease in the H_2 generation, whereas an increase was observed for the Gr wt. of 10%. This result is due to the changes in particles size of TiO_2. So, the addition of Gr did not cause significant changes in the BET specific area, and the reason for the enhanced photo-activity was not assigned to the changes of the BET specific area. The enhanced reactivity is attributed to the effective charge separation as well as the improved light absorption property.

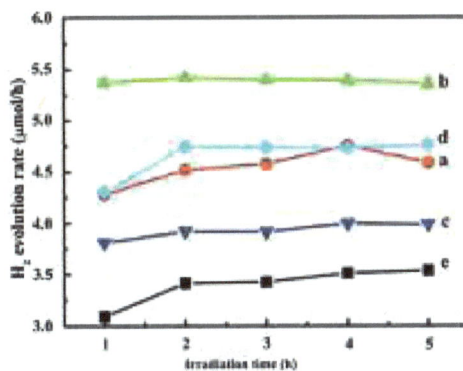

Fig. (13). Hydrogen evolution rate over TiO_2/Gr composites (a-d) and P25 (e) for comparsion. The content of GS is 0.8 (a), 2.0 (b), 5.0 (c), and 10.0 wt% (d), respectively [97].

The influence of both Gr content and the calcination temperature on the hydrogen generation was investigated by Zhang *et al.* [93]. So, these parameters can affect the photocatalytic activity of the obtained composites. Thus, the photocatalytic improvement reactivity of TiO_2/Gr materials is essentially due to the excellent electron conductivity of Gr by restraining the recombination of the (e^-/h^+) photogenerated. So, the photoexcited electron from the TiO_2 conduction band is transferred to the Gr, because the Fermi level of Gr is lower than the conduction band of TiO_2 (Fig. **14**) [98]. As well, the formation of chemically bonded TiO_2/Gr is also beneficial to light absorption in the visible range [99].

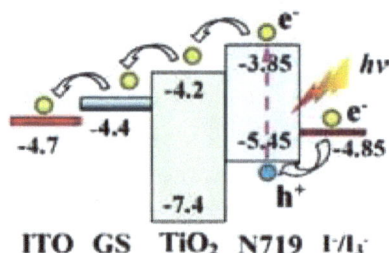

Fig. (14). Schematic energy-level diagram for the Gr/TiO_2 dye-sensitized solar cells (DSSCs) [98].

Thus, the effect of Gr content was also investigated. Zhang *et al.* prepared P25 TiO_2/Gr composite with different Gr content (0.25, 0.5, 1, 5, and 10 wt.%) using a solvothermal method [99]. The light source and scavenger used in the study were Xe lamp and methanol, respectively. P25/0.5%Gr nanocomposite showed the highest H_2 production during the whole photocatalytic reactions (668 mmol/h) compared to bare P25 (397 mmol/h). This is achieved by transferring the photoexcited electron from the conduction band of P25-TiO_2 to the Gr *via* a percolation mechanism [100]. Moreover, Gr consists of a two-dimensional π-conjugative structure; this structure makes Gr a competitive acceptor material [101].

Qian *et al.* used a microwave-hydrothermal method for the treatment of GO, while TiO_2 with (001) facets exposed were synthesized by the hydrothermal method to yield a Gr/TiO_2 nanosheet composite [102]. The activity of the composite was investigated using methanol as a sacrificial agent and under UV light. The influence of the Gr content in the composite on the photocatalytic activity shows that H_2 generated is increased with the increase of Gr up to 1 wt% (41 times higher than TiO_2), where a further increase in the Gr content led to a decrease in H_2 production (Fig. **15**). The enhancement photoreactivity is due to electron transport from TiO_2 to Gr. Also, it is confirmed by the comparison of the optimum composite with TiO_2/Degussa (P1.0) (Fig. **16**) [102]. The decrease in the photocatalytic activity with Gr content (greater than 1 wt.%) is due to the Gr loading shielding TiO_2 from the light *via* light scattering and opacity [103 - 107]. Furthermore, the authors proposed a photocatalytic H_2 production over the TiO_2/Gr composite mechanism, as illustrated in Fig. (**16**).

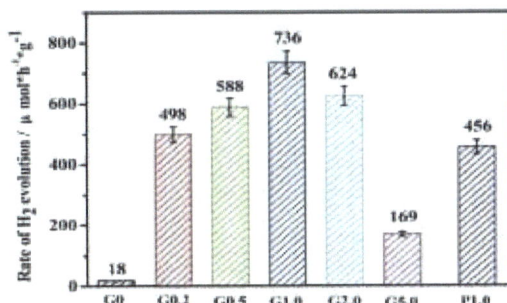

Fig. (15). H$_2$ generation as function of Gr content in Gr/TiO$_2$ photo-catalyst. G0, G0.2, G0.5, G1, G2.0, G5.0 represents 0, 0.2, 0.5, 1.0, 2.0, 5.0 Gr wt%. P1.0 is 1.0 wt% graphene/Degussa P25 composite [102].

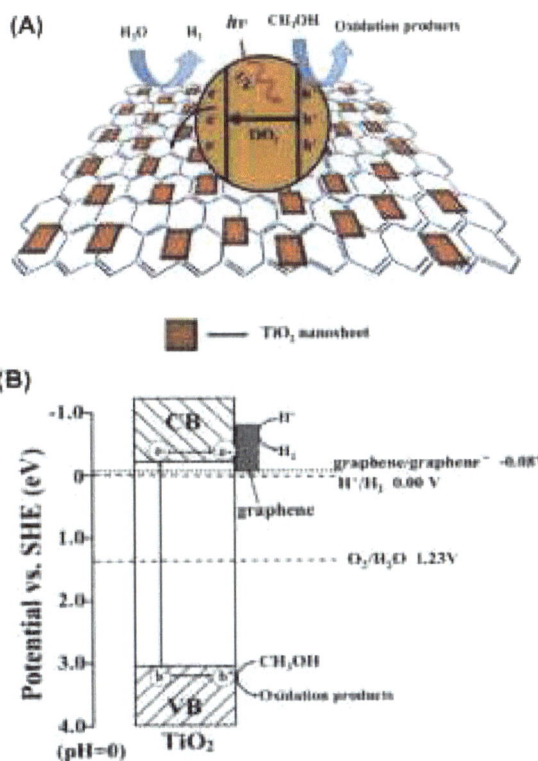

Fig. (16). (A) Illustration of charge movement and separation over the Gr/TiO$_2$ composite, (B) proposed photocatalytic mechanism for H$_2$ production *via* photocatalysis over Gr/TiO$_2$ composite [103].

The photo-generated electrons can migrate from the conduction bands of semiconductors to Gr, which can efficiently suppress the recombination of

Photoinduced carriers and promote photocatalytic hydrogen generation [108]. On the other hand, the extremely high conductivity of Gr can make the accepted electrons migrate rapidly across its 2D plane to active sites for hydrogen generation [109]. Thus, the role of Gr as an electron acceptor and transporter has been classified by various photocatalytic experiments. Table **1** depicts the most photocatalysts based on TiO_2-Gr used for the water splitting as well as the synthesis method, the reaction conditions, and their hydrogen amount generated *via* the process.

Table 1. Application of TiO_2-Gr for photocatalytic water splitting.

Materials	Synthesis method	Reaction conditions	Hydrogen generation	Ref
TiO_{2-x}/T (x = 0, 0.1, 0.5 and 1.T: calcination temperature)	Precipitation	Microsolar 300 Xenon lamp with a UV light(52 mW/cm^2) Photocatalyst:100 mg V: 100 mL sacrificial agents: ethanol. temperature:20° C	TiO_2-DC: 0.077 µmol h^{-1} g^{-1} TiO_2-0/650: 0.077 µmol h^{-1} g^{-1} TiO_2-CC:0.22 µmol h^{-1} g^{-1} TiO_2-1/725: 0.75 µmol h^{-1} g^{-1} TiO_2-0.1/675: 1,18µmol h^{-1} g^{-1}. TiO_2-0,5/700: 1.3 µmol h^{-1} g^{-1}	[110]
CNT-GR-TiO_2	Hydrolysis	Photocatalyst: 5 mg V: 50 mL (10 vol% of methanol) Xe–350 W lamp	TiO_2 9 µmol h^{-1} g^{-1} GR-TiO_2 22 µmol h^{-1} g^{-1} CNT-GRTiO$_2$ 29 µmol h^{-1} g^{-1}	[111]
P25-GR	Solvothermale	Photocatalyst: 100 mg V: 50 mL (45 mL H$_2$O 15 mL methanol) 300 W Xe lamp	P25 260 µmol h^{-1} P25-0.5%GR 330 µmol h^{-1}	[112]
P25-rGO-Co	hydrolysis-hydrothermal method	250-W high pressure mercury lamp photocatalyst:50 mg V: 80 mL (20% (v /v) CH$_3$OH)	P25: 20µmol h^{-1} P25–rGO–Co: 740 µmol h^{-1} P25–Co: 450µmol h^{-1} P25–rGO: 40µmol h^{-1}	[113]

(Table 1) cont.....

Materials	Synthesis method	Reaction conditions	Hydrogen generation	Ref
graphene–TiO$_2$ (TIO-9)		Photocatalyst: 50 mg V: methanol 50% Irradiation: 400-W high-pressure mercury lamp	TIO-9 (50mg) 0.45µmol/g·h TIO-9 (5mg) 0.95µmol/g·h	[114]

TiO$_2$-reduced Graphene Photocatalysts

TiO$_2$/Gr sheets were prepared by a one-step hydrothermal method [115]. The reduction of GO and the formation of TiO$_2$ were carried out simultaneously, leading to the well dispersion of generated TiO$_2$ nanoparticles on the surface of Gr sheets. Fan and co-authors elaborated TiO$_2$ nanocomposites and rGO by UV-assisted photocatalytic reduction, hydrazine reduction, and hydrothermal method, respectively [116]. These reduction techniques incorporate rGO into TiO$_2$, to varying degrees, significantly enhancing the photocatalytic activity of TiO$_2$ for H$_2$ generation.

The incorporation of rGO into commercial TiO$_2$ to enhance the H$_2$ generation was also investigated. Fan *et al.*, synthetized P25/rGO by the hydrothermal method where the enhancement photoreactivity was due to the formation of the intimate heterojunction [116]. Whereas Cui and co-authors investigated the H$_2$ generation using TiO$_2$-rGO photocatalyst prepared by a sol-gel method using Na$_2$S and Na$_2$SO$_3$ as sacrificial agents [117]. The H$_2$ produced over TiO$_2$-rGO composite (8.6 µmol h^{-1}) was twice that of P25 alone (4.5 µmol h^{-1}). The better of TiO$_2$/rGO efficiency compared to TiO$_2$ was also confirmed by Shen *et al.* [118]. So, the results demonstrated that under UV irradiation, TiO$_2$/rGO composite exhibited an H$_2$ generation rate of 4.0 m mol h^{-1} superior to TiO$_2$ nanoparticles. The improved activity of TiO$_2$/rGO is due to the electron transfer through the interface from TiO$_2$ to Gr. This transfer is facilitated by the difference in the energy levels between the TiO$_2$ and Gr by preventing the (e$^-$/h$^+$) charge recombination [119, 120]. Another study used methanol as a sacrifice agent was investigated [116, 121]. The optimum TiO$_2$/rGO ratio was 1/0.2 and methanol is the most efficient sacrifice agent in comparison with ethanol and isopropanol. Wang *et al.*, demonstrated that, 2 wt% rGO doped rGO/Pt-TiO$_2$ nanocomposites showed superior solar-driven H$_2$ generation (1075.68 µmol h^{-1} g^{-1}), which was 81 times and 5 times higher than bare TiO$_2$ and Pt/TiO$_2$ samples, respectively [122].

Khalid *et al.* prepared a visible light-responsive Gr-Fe/TiO$_2$ composite photocatalyst [123]. Due to the synergistic effect of the components in Gr-Fe/TiO$_2$ composite, Gr-Fe/TiO$_2$ showed more efficient charge separation than both Gr/TiO$_2$ and pure P25. Xiang *et al.* prepared a composite material, TiO$_2$

nanocrystals and layered MoS_2/Gr hybrid. It was found that the positive synergetic effect between the MoS_2 and Gr components in this hybrid co-catalyst can significantly enhance the photoactivity of TiO_2 [124].

Fig. (**17**) illustrated the proposed mechanism for the enhanced electron transfer in the TiO_2/MG system under irradiation assumes that the photoexcited electrons are transferred from the CB of TiO_2 not only to the MoS_2 nanosheets but also to the C atoms in the Gr sheets, which can effectively reduce H^+ to produce H_2 [124]. Rose *et al.* have prepared rGO/$BiVO_4$-Ru/$SrTiO_3$:Rh for Z-scheme photocatalytic water splitting under visible light irradiation [125]. This work demonstrated that rGO synthetized using $BiVO_4$ can be used as a solid-state electron mediator for a Z-scheme photocatalytic water splitting system (Fig. **18**). So, GO photo reduced by $BiVO_4$ is miscible in water and provides a great improvement in the activity for water splitting by efficiently transferring photoexcited electrons from the O_2 photocatalyst to the H_2 photocatalyst.

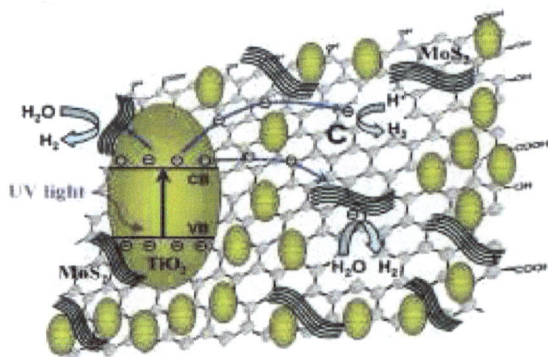

Fig. (17). Schematic illustration of the charge transfer in TiO_2/MG composites [124].

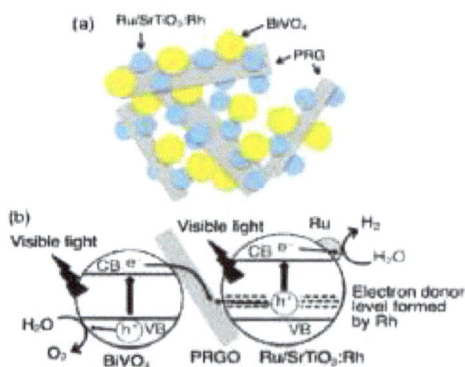

Fig. (18). (a) Schematic image of a suspension of Ru/$SrTiO_3$ and PRGO/$BiVO_4$ in water at pH 3.5. (b) Mechanism of water splitting in a Z-scheme photocatalysis system consisting of Ru/$SrTiO_3$:Rh and PRGO/$BiVO_4$ under visible-light irradiation [125].

ZnO-Graphene Photocatalysts

Compared to TiO$_2$, ZnO is largely used due to its higher electron mobility and a lower rate of carriers combination [126]. Also, ZnO with differents morphology can be easily synthetized. ZnO with gap energy of 3.37 eV, their photoactivity is observed only under UV irradiation light, the (e$^-$/h$^+$) photo-generated recombine easily on the surface lead to a poor utilization of photogenerated carriers [127]. ZnO/rGO hierarchical porous thin films show high photoelectrochemical water splitting performance [128]. A few approaches have been established for the elaboration of ZnO/GO photoanodes for PEC water splitting. So, triangular-shaped ZnO on GO demonstrates an efficient photoelectrochemical activity [129]. ZnO NWAs/rGO synthetized by sonochemical technique illustrate amelioration in the photo-electrochemical hydrogen production [130]. Some results of ZnO/GO are regrouped in Table **2**.

Table 2. Application of ZnO-graphene for photocatalytic water splitting.

Materials	Synthesis method	Reaction conditions	Hydrogen generation	Ref
ZnO/N-rGO	Solvothermal	photocatalyst: 100 mg V: 100 mL (80 mLH$_2$O/20 mL CH$_3$OH) Irradiation: 300 W Xe-lamp	ZnO/rGO 13.7 μmol.h^{-1} ZnO/N-rGO: 7 μmol.h^{-1} N-rGO:4.1 μmol/h^{-1}	[131]
ZrO$_2$-TiO$_2$ /Rgo	Facile chemical	photocatalyst: 5 mg Glycerol (5 vol%) Irradiation: UV	ZrO$_2$ 2000 μmolh^{-1}g^{-1} ZrO$_2$/rGO 2200 μmolh^{-1}g^{-1} ZrO$_2$-TiO$_2$/rGO 7773 μmolh^{-1}g^{-1}	[132]
ZnO/Rgo	microwave-assisted solvothermal	photocatalyst: 50 mg V:100 mL (Na$_2$S, 0.35 mol/L/Na$_2$SO$_3$, 0.25 mol/L), irradiation: 300 W Xenon lamp	ZnO:279.4 μmol/h/g. ZrG-0.2: 427.1μmol/h/g. ZrG-0.1: 499.5μmol/h/g. ZrG-1: 648.1μmol/h/g. ZrG-2:518.8μmol/h/g	[133]
RGO/ZnO		photocatalyst: 100 mg V: 60 mL (Na2S, 0.1 mol/L/Na$_2$SO$_3$, 0.05 mol/L) irradiation: UV-light	ZnO/RGO: 289 μmol/g ZnO:61.5 μmol/g	[134]
ZnO-ZnS/ Graphene		photocatalyst:50 mg of V:100 mL glycerol (1, 10, 20, 30, 40, 50 vol %). irradiation: 300W high-pressure mercury lamp temperature: 30°C	ZnO-ZnS/Gr 1070 μmol h^{-1} g^{-1} ZnO-ZnS 193 μmol h^{-1} g^{-1}	[135]

CONCLUSION

The graphene-based materials gained huge interest in the last decades due to the cost-effective, and easy way to elaborate both graphene and graphene oxide materials. Recently, many efforts have been made to develop graphene-based materials properties in various fields. Thus, these materials are largely synthesized by the destruction and construction method. Owing to the two-dimensional π - conjugation structure, graphene served as an acceptor of the photogenerated electrons of TiO_2 and transporter to separate the photogenerated pairs (e^-/h^+) efficiently.

Graphene-based materials possessed enhanced light absorption ability and a lower recombination rate of photogenerated pairs (e^-/h^+), therefore displaying higher photocatalytic activity toward the hydrogen generation. Through the examples presented this chapter, it is clearly demonstrating the high potential applications of the graphene-based materials clean and renewable energy.

CONSENT FOR PUBLICATION

Not applicable.

CONFLICT OF INTEREST

The author declares no conflict of interest, financial or otherwise.

ACKNOWLEDGEMENTS

This work is supported by the Directorate-General for Scientific Research and Technological Development DGRSDT, and they are thanked for their financial support.

REFERENCES

[1] Chowdhury, S.; Balasubramanian, R. Recent advances in the use of graphene-family nanoadsorbents for removal of toxic pollutants from wastewater. *Adv. Colloid Interface Sci.,* **2014**, *204*, 35-56.
[http://dx.doi.org/10.1016/j.cis.2013.12.005] [PMID: 24412086]

[2] Cardinali, M.; Valentini, L.; Fabbri, P.; Kenny, J.M. Radio frequency plasma assisted exfoliation and reduction of large-area graphene oxide platelets produced by a mechanical transfer process. *Chem. Phys. Lett.,* **2011**, *508*(4–6), 285-288.
[http://dx.doi.org/10.1016/j.cplett.2011.04.065]

[3] Bhuyan, M.S.A.; Uddin, M.N.; Islam, M.M.; Bipasha, F.A.; Hossain, S.S. Synthesis of graphene. *Int. Nano Lett.,* **2016**, *6*, 65-83.
[http://dx.doi.org/10.1007/s40089-015-0176-1]

[4] Shekhawat, A.; Ritchie, R.O. Toughness and strength of nanocrystalline graphene. *Nat. Commun.,* **2016**, *7*, 10546.
[http://dx.doi.org/10.1038/ncomms10546] [PMID: 26817712]

[5]   Kuila, T.; Bose, S.; Mishra, A.K.; Khanra, P.; Kim, N.H.; Lee, J.H. Chemical functionalization of graphene and its applications. *Prog. Mater. Sci.,* **2012**, *57*(7), 1061-1105.
[http://dx.doi.org/10.1016/j.pmatsci.2012.03.002]

[6]   Saba, N.; Jawaid, M. *4 - Energy and environmental applications of graphene and its derivatives*; Polymer-based Nanocomposites for Energy and Environmental Applications, **2018**, pp. 105-129.
[http://dx.doi.org/10.1016/B978-0-08-102262-7.00004-0]

[7]   Liu, J.; Cui, L.; Losic, D. Graphene and graphene oxide as new nanocarriers for drug delivery applications. *Acta Biomater.,* **2013**, *9*(12), 9243-9257.
[http://dx.doi.org/10.1016/j.actbio.2013.08.016] [PMID: 23958782]

[8]   Pumera, M. Graphene-based nanomaterials for energy storage. *Energy Environ. Sci.,* **2011**, *4*, 668-674.
[http://dx.doi.org/10.1039/C0EE00295J]

[9]   Wang, S.; Sun, H.; Ang, H.M.; Tadé, M.O. Adsorptive remediation of environ-mental pollutants using novel graphene-based nanomaterials. *Chem. Eng. J.,* **2013**, *226*, 336-347.
[http://dx.doi.org/10.1016/j.cej.2013.04.070]

[10]   Geim, A.K.; Novoselov, K.S. The rise of graphene. *Nat. Mater.,* **2007**, *6*(3), 183-191.
[http://dx.doi.org/10.1038/nmat1849] [PMID: 17330084]

[11]   Novoselov, K.S.; Jiang, D.; Schedin, F.; Booth, T.J.; Khotkevich, V.V.; Morozov, S.V.; Geim, A.K. Two-dimensional atomic crystals. *Proc. Natl. Acad. Sci. USA,* **2005**, *102*(30), 10451-10453.
[http://dx.doi.org/10.1073/pnas.0502848102] [PMID: 16027370]

[12]   Novoselov, K.S.; Geim, A.K.; Morozov, S.V.; Jiang, D.; Zhang, Y.; Dubonos, S.V.; Grigorieva, I.V.; Firsov, A.A. Electric field effect in atomically thin carbon films. *Science,* **2004**, *306*(5696), 666-669.
[http://dx.doi.org/10.1126/science.1102896] [PMID: 15499015]

[13]   Zhang, Y.; Tan, Y.W.; Stormer, H.L.; Kim, P. Experimental observation of the quantum Hall effect and Berry's phase in graphene. *Nature,* **2005**, *438*(7065), 201-204.
[http://dx.doi.org/10.1038/nature04235] [PMID: 16281031]

[14]   Geim, A.K. Graphene: status and prospects. *Science,* **2009**, *324*(5934), 1530-1534.
[http://dx.doi.org/10.1126/science.1158877] [PMID: 19541989]

[15]   Rao, C.N.R.; Sood, A.K.; Subrahmanyam, K.S.; Govindaraj, A. Graphene: the new two-dimensional nanomaterial. *Angew. Chem. Int. Ed. Engl.,* **2009**, *48*(42), 7752-7777.
[http://dx.doi.org/10.1002/anie.200901678] [PMID: 19784976]

[16]   Zhu, Y.; Murali, S.; Cai, W.; Li, X.; Suk, J.W.; Potts, J.R.; Ruoff, R.S. Graphene and graphene oxide: synthesis, properties, and applications. *Adv. Mater.,* **2010**, *22*(35), 3906-3924.
[http://dx.doi.org/10.1002/adma.201001068] [PMID: 20706983]

[17]   Du, X.; Skachko, I.; Barker, A.; Andrei, E.Y. Approaching ballistic transport in suspended graphene. *Nat. Nanotechnol.,* **2008**, *3*(8), 491-495.
[http://dx.doi.org/10.1038/nnano.2008.199] [PMID: 18685637]

[18]   Balandin, A.A.; Ghosh, S.; Bao, W.; Calizo, I.; Teweldebrhan, D.; Miao, F.; Lau, C.N. Superior thermal conductivity of single-layer graphene. *Nano Lett.,* **2008**, *8*(3), 902-907.
[http://dx.doi.org/10.1021/nl0731872] [PMID: 18284217]

[19]   Grantab, R.; Shenoy, V.B.; Ruoff, R.S. Anomalous strength characteristics of tilt grain boundaries in graphene. *Science,* **2010**, *330*(6006), 946-948.
[http://dx.doi.org/10.1126/science.1196893] [PMID: 21071664]

[20]   Park, S.; Ruoff, R.S. Chemical methods for the production of graphenes. *Nat. Nanotechnol.,* **2009**, *4*(4), 217-224.
[http://dx.doi.org/10.1038/nnano.2009.58] [PMID: 19350030]

[21]   Pei, S.F.; Zhao, J.P.; Du, J.H.; Ren, W.C.; Cheng, H.M. Derect reduction of graphene oxide films into highly conductive and flexible graphene films by hydrohalic acids. *Carbon,* **2010**, *48*, 4466-4474.

[http://dx.doi.org/10.1016/j.carbon.2010.08.006]

[22]   Allen, M.J.; Tung, V.C.; Kaner, R.B. Honeycomb carbon: a review of graphene. *Chem. Rev.,* **2010**, *110*(1), 132-145.
[http://dx.doi.org/10.1021/cr900070d] [PMID: 19610631]

[23]   Vickery, J.L.; Patil, A.J.; Mann, S. Fabrication of graphene–polymer nanocomposites with giger-order three-dimensional architectures. *Adv. Mater.,* **2009**, *21*, 2180-2184.
[http://dx.doi.org/10.1002/adma.200803606]

[24]   Wang, H.W.; Hu, Z.A.; Chang, Y.Q.; Chen, Y.L.; Lei, Z.Q.; Zhang, Z.Y. Facile solvothermal synthesis of a graphee nanosheet-bismuth oxide composite and its electrochemical characteristics. *Electrochim. Acta,* **2010**, *55*, 8974-8980.
[http://dx.doi.org/10.1016/j.electacta.2010.08.048]

[25]   Sun, S.R.; Gao, L.; Liu, Y.Q. Enhanced dye-sensitized solar cell using graphene-TiO$_2$ photoanode prepared by heterogeneous coagulation. *Appl. Phys. Lett.,* **2010**, *96*, 083113.
[http://dx.doi.org/10.1063/1.3318466]

[26]   Ehrenfreund, P.; Foing, B.H. Astronomy. Fullerenes and cosmic carbon. *Science,* **2010**, *329*(5996), 1159-1160.
[http://dx.doi.org/10.1126/science.1194855] [PMID: 20813945]

[27]   Novoselov, K. S.; Geim, A. K.; Morozov, S. V.; Jiang, D.; Katsnelson, M. I.; Grigorieva, I. V.; Dubonos, S. V.; Firsov, A. A. Two-dimensional gas of mass less Dirac fermions in graphene. *Nature,* *438*, 197-9.

[28]   Pesin, D.; MacDonald, A. H. Spintronics and pseudospintronics in graphene and topological insulators. *Nature mater,* **2012**, *11*(5), 409.
[http://dx.doi.org/10.1038/nmat3305]

[29]   Scheike, T.; Böhlmann, W.; Esquinazi, P.; Barzola-Quiquia, J.; Ballestar, A.; Setzer, A. Can doping graphite trigger room temperature superconductivity? Evidence for granular high-temperature superconductivity in water-treated graphite powder. *Adv. Mater.,* **2012**, *24*(43), 5826-5831.
[http://dx.doi.org/10.1002/adma.201202219] [PMID: 22949348]

[30]   Dasari, B.L.; Nouri, J.M.; Brabazon, D.; Naher, S. Graphene and derivatives-Synthesis techniques, properties and their energy applications. *Energy,* **2017**, *140*, 766-778.
[http://dx.doi.org/10.1016/j.energy.2017.08.048]

[31]   Ren, S.; Rong, P.; Yu, Q. Preparations, properties and applications of graphene in functional devices: a concise review. *Ceram. Int.,* **2018**, *44*(11), 11940-11955.
[http://dx.doi.org/10.1016/j.ceramint.2018.04.089]

[32]   Yu, H.; Zhang, B.; Bulin, C.; Li, R.; Xing, R. High-efficient synthesis of graphene oxide based on improved hummers method. *Sci. Rep.,* **2016**, *6*, 36143.
[http://dx.doi.org/10.1038/srep36143] [PMID: 27808164]

[33]   Zhong, Y.L.; Tian, Z.; Simon, G.P.; Li, D. Scalable production of graphene *via* wet chemistry: progress and challenges. *Mater. Today,* **2015**, *18*(2), 73-78.
[http://dx.doi.org/10.1016/j.mattod.2014.08.019]

[34]   Edwards, R.S.; Coleman, K.S. Graphene synthesis: relationship to applications. *Nanoscale,* **2013**, *5*(1), 38-51.
[http://dx.doi.org/10.1039/C2NR32629A] [PMID: 23160190]

[35]   Ambrosi, A.; Chua, C.K.; Bonanni, A.; Pumera, M. Electrochemistry of graphene and related materials. *Chem. Rev.,* **2014**, *114*(14), 7150-7188.
[http://dx.doi.org/10.1021/cr500023c] [PMID: 24895834]

[36]   Novoselov, K.S.; Geim, A.K.; Morozov, S.V.; Jiang, D.; Zhang, Y.; Dubonos, S.V.; Grigorieva, I.V.; Firsov, A.A. Electric field effect in atomically thin carbon films. *Science,* **2004**, *306*(5696), 666-669.
[http://dx.doi.org/10.1126/science.1102896] [PMID: 15499015]

[37]   Chen, Y.; Zhao, H.; Sheng, L.; Yu, L.; An, K.; Xu, J.; Ando, Y.; Zhao, X. Mass-production of highlycrystalline few-layer graphene sheets by arc discharge in various H2-inert gas mixtures. *Chem. Phys. Lett.,* **2012**, *538*, 72-76.
[http://dx.doi.org/10.1016/j.cplett.2012.04.020]

[38]   Emiru, T.F.; Ayele, D.W. Controlled synthesis, characterization and reduction of graphene oxide: a convenient method for large scale production. Egypt. *J. Basic Appl. Sci.,* **2017**, *4*, 74-79.

[39]   Güler, Ö.; Güler, S.H.; Selen, V.; Albayrak, M.G.; Evin, E. Production of graphene layer by liquid-phase exfoliation with low sonication power and sonication time from synthesized expanded graphite. *Fuller. Nanotub. Carbon Nanostruct.,* **2016**, *24*, 123-127.
[http://dx.doi.org/10.1080/1536383X.2015.1114472]

[40]   Valentini, L. Formation of unzipped carbon nanotubes by CF4 plasma treatment. *Diamond Related Materials,* **2011**, *20*, 445-448.
[http://dx.doi.org/10.1016/j.diamond.2011.01.038]

[41]   Wang, X.; You, H.; Liu, F.; Li, M.; Wan, L.; Li, S.; Li, Q.; Xu, Y.; Tian, R.; Yu, Z.; Xiang, D.; Cheng, J. Large-scale synthesis of few-layered graphene using CVD. *J. Chem. Vapor Deposition,* **2009**, *15*(1–3), 53-56.
[http://dx.doi.org/10.1002/cvde.200806737]

[42]   Chae, S.J. G€unes, F., Kim, K.K., Kim, E.S., Han, G.H., Kim, S.M., Shin, H.-J., Yoon, S.-M., Choi, J.-Y., Park, M.H., Yang, C.W., Pribat, D., Lee, Y.H.: Synthesis of large-area graphene layers on poly-nickel substrate by chemical vapor deposition: wrinkle formation. *Adv. Mater.,* **2009**, *21*(22), 2328-2333.
[http://dx.doi.org/10.1002/adma.200803016]

[43]   Yang, Y.; Liu, R.; Wu, J.; Jiang, X.; Cao, P.; Hu, X.; Pan, T.; Qiu, C.; Yang, J.; Song, Y.; Wu, D.; Su, Y. Bottom-up fabrication of graphene on silicon/silica substrate *via* a facile soft-hard template approach. *Sci. Rep.,* **2015**, *5*, 13480.
[http://dx.doi.org/10.1038/srep13480] [PMID: 26311022]

[44]   Albert, D.; Michael, F. Substrate-free microwave synthesis of graphene: experimental conditions and hydrocarbon precursors. *New J. Phys.,* **2010**, *12*, 125013.
[http://dx.doi.org/10.1088/1367-2630/12/12/125013]

[45]   Yazdi, G.; Iakimov, T.; Yakimova, R. Epitaxial graphene on SiC: a review of growth and characterization. *Crystal,* **2016**, *6*, 53.
[http://dx.doi.org/10.3390/cryst6050053]

[46]   Singh, V.; Joung, D.; Zhai, L.; Das, S.; Khondaker, S.I.; Seal, S. Graphene based mate- rials: past, present and future. *Prog. Mater. Sci.,* **2011**, *56*, 1178-1271.
[http://dx.doi.org/10.1016/j.pmatsci.2011.03.003]

[47]   Hu, M.; Yao, Z.; Wang, X. Characterization techniques for graphene-based materials in catalysis. *AIMS Mater. Sci.,* **2017**, *4*(3), 755-788. [J].
[http://dx.doi.org/10.3934/matersci.2017.3.755]

[48]   Ren, P-G.; Yan, D-X.; Ji, X.; Chen, T.; Li, Z-M. Temperature dependence of graphene oxide reduced by hydrazine hydrate. *Nanotechnology,* **2011**, *22*(5), 055705.
[http://dx.doi.org/10.1088/0957-4484/22/5/055705] [PMID: 21178230]

[49]   Tripathi, P.; Prakash, C.; Shaz, M.; Srivastava, O.N. Synthesis of High-Quality Graphene through Electrochemical Exfoliation of Graphite in Alkaline Electrolyte. **2013**.

[50]   Yoon, D.; Moonand, H.; Cheong, H. Variations in the Raman Spectrum as a Function of the Number of Graphene Layers. *J. Korean Phys. Soc.,* **2009**, *55*, 1299-1303.
[http://dx.doi.org/10.3938/jkps.55.1299]

[51]   Choi, W.; Lahiri, I.; Seelaboyina, R.; Kang, Y.S. *Synthesis of Graphene and Its Applications: A Review, Critical Reviews in Solid State and Materials Sciences*; Taylor&Francis, **2010**.

[52] Li, W.; Tan, C.; Lowe, M.A.; Abruña, H.D.; Ralph, D.C. Electrochemistry of individual monolayer graphene sheets. *ACS Nano,* **2011**, *5*(3), 2264-2270.
[http://dx.doi.org/10.1021/nn103537q] [PMID: 21332139]

[53] Lancelot, E. Nanotechnology, Perspectives on Raman spectroscopy of Graphene. **2011**.

[54] Lee, B-J. Jeong, G. - H. Comparative study on graphene growth mechanism using Ni films, Ni/Mo sheets, and Pt substrates. *Appl. Phys., A Mater. Sci. Process.,* **2014**, *116*, 15-24.
[http://dx.doi.org/10.1007/s00339-014-8493-1]

[55] You, Y.; Lakshmi, V. N.; Sinha, S. K.; Haven, W. AFM characterization of Multilayered Graphene film used as Hydrogen Sensor. *ASEE 2014ZoIConf,* **2014**, 5-7.

[56] Liu, W-W.; Chai, S-P.; Mohamed, A.R.; Hashim, U. Synthesis and characterization of graphene and carbon nanotubes: a review on the past and recent develop ments. *J. Ind. Eng. Chem.,* **2014**, *20*(4), 1171-1185.
[http://dx.doi.org/10.1016/j.jiec.2013.08.028]

[57] Dong, L-X.; Chen, Q. Properties, synthesis, and characterization of graphene. *Front. Mater. Sci.,* **2010**, *4*(1), 45-51.
[http://dx.doi.org/10.1007/s11706-010-0014-3]

[58] Tu, Z.; Liu, Z.; Li, Y.; Yang, F.; Zhang, L.; Zhao, Z. Xu, C., Wu, S., Liu, H., Yang, H., Richard, P. Controllable growth of 1–7 layers of graphene by chemical vapour deposition. *Carbon,* **2014**, *73*, 252-258.
[http://dx.doi.org/10.1016/j.carbon.2014.02.061]

[59] Banhart, F.; Kotakoski, J.; Krasheninnikov, A.V. Structural defects in graphene. *ACS Nano,* **2011**, *5*(1), 26-41.
[http://dx.doi.org/10.1021/nn102598m] [PMID: 21090760]

[60] Lee, H.C.; Liu, W-W.; Chai, S-P.; Mohamed, A.R.; Aziz, A.; Khe, C-S. Hi- dayah, N.M.S., Hashim, U. Review of the synthesis, transfer, characterization and growth mechanisms of single and multilayer graphene. *RSC Advances,* **2017**, *7*(26), 15644-15693.
[http://dx.doi.org/10.1039/C7RA00392G]

[61] Suk, J.J.; Gyu, K.H.; Sung, L.J. Heterojunction semiconductors: a strategy to develop efficient photocatalytic materials for visible light water splitting. *Catal. Today,* **2012**, *185*, 270-277.
[http://dx.doi.org/10.1016/j.cattod.2011.07.008]

[62] Fujishima, A.; Honda, K. Electrochemical photolysis of water at a semiconductor electrode. *Nature,* **1972**, *238*(5358), 37-38.
[http://dx.doi.org/10.1038/238037a0] [PMID: 12635268]

[63] Kudo, A.; Miseki, Y. Heterogeneous photocatalyst materials for water splitting. *Chem. Soc. Rev.,* **2009**, *38*(1), 253-278.
[http://dx.doi.org/10.1039/B800489G] [PMID: 19088977]

[64] Idriss, H. The illusive photocatalytic water splitting reaction using sun light on suspended nanoparticles. Is there a way forward? In: *Catal. Sci. Techn*; , **2019**.
[http://dx.doi.org/10.1039/C9CY01818B]

[65] Maeda, K.; Domen, K. Photocatalytic Water Splitting: Recent Progress and Future Challenges. *J. Phys. Chem. Lett.,* **2010**, *1*(18), 2655-2661.
[http://dx.doi.org/10.1021/jz1007966]

[66] Sayama, K.; Mukasa, K.; Abe, R.; Abe, Y.; Arakawa, H. *Chem. Commun.,* **2001**, *23*, 2416-2417.
[http://dx.doi.org/10.1039/b107673f]

[67] Hisatomi, T.; Kubota, J.; Domen, K. Recent advances in semiconductors for photocatalytic and photoelectrochemical water splitting. *Chem. Soc. Rev.,* **2014**, *43*(22), 7520-7535.
[http://dx.doi.org/10.1039/C3CS60378D] [PMID: 24413305]

[68]   Xiang, Q.; Cheng, B.; Yu, J. Graphene-based photocatalysts for solar-fuel generation. *Angew. Chem. Int. Ed. Engl.,* **2015**, *54*(39), 11350-11366.
[http://dx.doi.org/10.1002/anie.201411096] [PMID: 26079429]

[69]   Abea, R. Recent progress on photocatalytic and photoelectrochemical water splitting under visible light irradiation. *J. Photochem. Photobiol. Chem.,* **2010**, *11*, 179-209.
[http://dx.doi.org/10.1016/j.jphotochemrev.2011.02.003]

[70]   Boudjemaa, A.; Trari, M.; Bachari, K. Photo-electrochemical properties of an active site working on Fe-SBA-15 catalyst: application to the water reduction under visible light irradiation. *Environ. Prog. Sustain. Energy,* **2014**, *33*, 141-146.
[http://dx.doi.org/10.1002/ep.11766]

[71]   Boudjemaa, A.; Mokrani, T.; Bachari, K.; Coville, N.J. Electrochemical and photo-electrochemical properties of carbon spheres prepared *via* chemical vapor deposition. *Mater. Sci. Semicond. Process.,* **2015**, *30*, 456-461.
[http://dx.doi.org/10.1016/j.mssp.2014.10.050]

[72]   Abe, R. Recent progress on photocatalytic and photoelectrochemical water splitting under visible light irradiation. J. Photochem.Photob. C. *Phytochem. Rev.,* **2010**, *11*(4), 179-209.
[http://dx.doi.org/10.1016/j.jphotochemrev.2011.02.003]

[73]   Boudjemaa, A.; Bouarab, R.; Saadi, S.; Bouguelia, A.; Trari, M. Photo electrochemical H2-generation over Spinel FeCr2O4 in X2- solutions (X2- = S2- and SO32-). *Appl. Energy,* **2009**, *86*, 1080-1086.
[http://dx.doi.org/10.1016/j.apenergy.2008.06.007]

[74]   Perreault, F.; Fonseca de Faria, A.; Elimelech, M. Environmental applications of graphene-based nanomaterials. *Chem. Soc. Rev.,* **2015**, *44*(16), 5861-5896.
[http://dx.doi.org/10.1039/C5CS00021A] [PMID: 25812036]

[75]   Yang, M.Q.; Xu, Y.J. Photocatalytic conversion of CO_2 over graphene-based composites: current status and future perspective. *Nanoscale Horiz.,* **2016**, *1*(3), 185-200.
[http://dx.doi.org/10.1039/C5NH00113G] [PMID: 32260621]

[76]   Kowalska, E.; Abe, R.; Ohtani, B. Visible light-induced photocatalytic reaction of gold-modified titanium(IV) oxide particles: action spectrum analysis. *Chem. Commun. (Camb.),* **2009**, (2), 241-243.
[http://dx.doi.org/10.1039/B815679D] [PMID: 19099082]

[77]   Gopalakrishnan, M.; Gopalakrishnan, S.; Bhalerao, G.M.; Jeganathan, K. Multiband InGaN nanowires with enhanced visible photon absorption for efficient photoelectrochemical water splitting. *J. Power Sources,* **2017**, *337*, 130-136.
[http://dx.doi.org/10.1016/j.jpowsour.2016.10.099]

[78]   Boudjemaa, A.; Boumaza, S.; Trari, M.; Bouarab, R.; Bouguelia, A. Physical and photo-electrochemical characterization of α-Fe_2O_3. Application for hydrogen production. *Int. J. Hydrogen Energy,* **2009**, *34*, 4268-4274.
[http://dx.doi.org/10.1016/j.ijhydene.2009.03.044]

[79]   Boudjemaa, A.; Bachari, K. Trari. M. Photo-electrochemical Characterization of Porous Material Fe-FSM-16. Application for Hydrogen Production. *Mater. Sci. Semicond. Process.,* **2013**, *16*, 838-844.
[http://dx.doi.org/10.1016/j.mssp.2013.01.008]

[80]   Zazoua, H.; Boudjemaa, A.; Chebout, R.; Bachari, K. Enhanced photocatalytic hydrogen production under visible light over a material based on magnesium ferrite derived from layered double hydroxides (LDHs). *Int. J. Energy Res.,* **2014**, *38*, 2010-2018.
[http://dx.doi.org/10.1002/er.3215]

[81]   Boudjemaa, A.; Bachari, K.; Trari, M. Photo-induced hydrogen on iron hexagonal mesoporous silica (Fe-HMS) photo-catalyst. *Int. J. Energy Res.,* **2013**, *37*, 171-178.
[http://dx.doi.org/10.1002/er.1880]

[82]   Grantab, R.; Shenoy, V.B.; Ruoff, R.S. Anomalous strength characteristics of tilt grain boundaries in

graphene. *Science,* **2010**, *330*(6006), 946-948.
[http://dx.doi.org/10.1126/science.1196893] [PMID: 21071664]

[83]   Nowotny, J.; Bak, T.; Nowotny, M.K.; Sheppard, L.R. Titanium dioxide for solar-hydrogen I. Functional properties. *Int. J. Hydrogen Energy,* **2007**, *32*, 2609-2629.
[http://dx.doi.org/10.1016/j.ijhydene.2006.09.004]

[84]   Rico-Oller, B.; Boudjemaa, A.; Bahruji, H.; Kebir, M.; Prashar, S.; Bachari, K.; Fajardo, M.; Gómez-Ruiz, S. Photodegradation of organic pollutants in water and green hydrogen production via methanol photoreforming of doped titanium oxide nanoparticles. *Sci. Total Environ.,* **2016**, *563-564*, 921-932.
[http://dx.doi.org/10.1016/j.scitotenv.2015.10.101] [PMID: 26524993]

[85]   Zhang, Y.; Tang, Z.R.; Fu, X.; Xu, Y.J. TiO_2-graphene nanocomposites for gas-phase photocatalytic degradation of volatile aromatic pollutant: is TiO_2-graphene truly different from other TiO_2-carbon composite materials? *ACS Nano,* **2010**, *4*(12), 7303-7314.
[http://dx.doi.org/10.1021/nn1024219] [PMID: 21117654]

[86]   Liang, Y.; Wang, H.; Casalongue, H.S.; Chen, Z.; Dai, H. TiO_2 nanocrystals grown on graphene as advanced photocatalytic hybrid materials. *Nano Res.,* **2010**, *3*, 701-705.
[http://dx.doi.org/10.1007/s12274-010-0033-5]

[87]   Liang, Y.T.; Vijayan, B.K.; Gray, K.A.; Hersam, M.C. Indirect optical transitions in hybrid spheres with alternating layers of titania and graphene oxide nanosheets. *Nano Lett.,* **2011**, *11*, 2865-2870.
[http://dx.doi.org/10.1021/nl2012906] [PMID: 21688817]

[88]   Zhang, N.; Zhang, Y.H.; Pan, X.Y.; Fu, X.Z.; Liu, S.Q.; Xu, Y.J. Recent progress on graphene-based photocatalysts: current status and future perspectives. *J. Phys. Chem. C,* **2011**, *115*, 23501-23511.
[http://dx.doi.org/10.1021/jp208661n]

[89]   Xu, T.G.; Zhang, L.W.; Cheng, H.Y.; Zhu, Y.F. UV-assisted photocatalytic synthesis of TiO_2-reduced graphene oxide with enhanced photocatalytic activity in decomposition of sarin in gas phase. *Appl. Catal. B,* **2011**, *101*, 382-387.
[http://dx.doi.org/10.1016/j.apcatb.2010.10.007]

[90]   Liu, G.; Yin, L.C.; Wang, J.; Niu, P.; Zhen, C.; Xie, Y.; Cheng, H.M. A red anatase TiO_2 photocatalyst for solar energy conversion. *Energy Environ. Sci.,* **2012**, *5*, 9603-9610.
[http://dx.doi.org/10.1039/c2ee22930g]

[91]   Sang, Y.; Zhao, Z.; Tian, J.; Hao, P.; Jiang, H.; Liu, H.; Claverie, J.P. Enhanced photocatalytic property of reduced graphene oxide/TiO_2 nanobelt surface heterostructures constructed by an *in situ* photochemical reduction method. *Small,* **2014**, *10*(18), 3775-3782.
[http://dx.doi.org/10.1002/smll.201303489] [PMID: 24888721]

[92]   Xie, G.; Zhang, K.; Guo, B.; Liu, Q.; Fang, L.; Gong, J.R. Graphene-based materials for hydrogen generation from light-driven water splitting. *Adv. Mater.,* **2013**, *25*(28), 3820-3839.
[http://dx.doi.org/10.1002/adma.201301207] [PMID: 23813606]

[93]   Zhang, X.Y.; Li, H.P.; Cui, X.L.; Lin, Y.H. Graphene/TiO_2 nanocomposites: synthesis, characterization and application in hydrogen evolution from water photocatalytic splitting. *J. Mater. Chem.,* **2010**, *20*, 2801-2806.
[http://dx.doi.org/10.1039/b917240h]

[94]   Nethravathi, C.; Rajamathi, M. Chemically modified graphene sheets produced by the solvothermal reduction of colloidal dispersions of graphite oxide. *Carbon,* **2008**, *46*, 1994-1998.
[http://dx.doi.org/10.1016/j.carbon.2008.08.013]

[95]   Morales-Torres, S.; Pastrana-Martínez, L.M.; Figueiredo, J.L.; Faria, J.L.; Silva, A.M.T. Design of graphene-based TiO_2 photocatalysts--a review. *Environ. Sci. Pollut. Res. Int.,* **2012**, *19*(9), 3676-3687.
[http://dx.doi.org/10.1007/s11356-012-0939-4] [PMID: 22782794]

[96]   Wang, X.; Zhi, L.; Müllen, K. Transparent, conductive graphene electrodes for dye-sensitized solar cells. *Nano Lett.,* **2008**, *8*(1), 323-327.

[http://dx.doi.org/10.1021/nl072838r] [PMID: 18069877]

[97]   Zhang, X.; Sun, Y.; Cui, X.; Jiang, Z. A green and facile synthesis of TiO$_2$/graphene nanocomposites and their photocatalytic activity for hydrogen evolution. *Int. J. Hydrogen Energy,* **2012**, *37*, 811-815.
       [http://dx.doi.org/10.1016/j.ijhydene.2011.04.053]

[98]   Tang, Y-B.; Lee, C-S.; Xu, J.; Liu, Z-T.; Chen, Z-H.; He, Z.; Cao, Y-L.; Yuan, G.; Song, H.; Chen, L.; Luo, L.; Cheng, H-M.; Zhang, W-J.; Bello, I.; Lee, S-T. Incorporation of graphenes in nanostructured TiO(2) films *via* molecular grafting for dye-sensitized solar cell application. *ACS Nano,* **2010**, *4*(6), 3482-3488.
       [http://dx.doi.org/10.1021/nn100449w] [PMID: 20455548]

[99]   Zhang, H.; Lv, X.; Li, Y.; Wang, Y.; Li, J. P25-graphene composite as a high performance photocatalyst. *ACS Nano,* **2010**, *4*(1), 380-386.
       [http://dx.doi.org/10.1021/nn901221k] [PMID: 20041631]

[100] Cheng, P.; Yang, Z.; Wang, H.; Cheng, W.; Chen, M.; Shangguan, W.; Ding, G. TiO$_2$-graphene nanocomposites for photocatalytic hydrogen production from splitting water. *Int. J. Hydrogen Energy,* **2012**, *37*, 2224-22230.
       [http://dx.doi.org/10.1016/j.ijhydene.2011.11.004]

[101] Wang, X.; Zhi, L.; Müllen, K. Transparent, conductive graphene electrodes for dye-sensitized solar cells. *Nano Lett.,* **2008**, *8*(1), 323-327.
       [http://dx.doi.org/10.1021/nl072838r] [PMID: 18069877]

[102] Qian, L.; Zunfeng, L.; Xiaoyan, Z.; Liying, Y.; Nan, Z.; Guiling, P.; Shougen, Y.; Yongsheng, C.; Jun, W. Polymer photovoltaic cells based on solution-processable graphene and P3HT. *Adv. Funct. Mater.,* **2009**, *19*, 894-904.
       [http://dx.doi.org/10.1002/adfm.200800954]

[103] Xiang, Q.; Yu, J.; Jaroniec, M. Enhanced photocatalytic H$_2$-production activity of graphene-modified titania nanosheets. *Nanoscale,* **2011**, *3*(9), 3670-3678.
       [http://dx.doi.org/10.1039/c1nr10610d] [PMID: 21826308]

[104] Yu, J.; Hai, Y.; Jaroniec, M. Photocatalytic hydrogen production over CuO-modified titania. *J. Colloid Interface Sci.,* **2011**, *357*(1), 223-228.
       [http://dx.doi.org/10.1016/j.jcis.2011.01.101] [PMID: 21345445]

[105] Yu, J.; Hai, Y.; Cheng, B. Enhanced photocatalytic H2-production activity of TiO$_2$ by Ni(OH)2 cluster modification. *J. Phys. Chem. C,* **2011**, *115*, 4953-4968.
       [http://dx.doi.org/10.1021/jp111562d]

[106] Zhang, H.; Xu, P.; Du, G.; Chen, Z.; Oh, K.; Pan, D.; Jiao, Z. A facile one-step synthesis of TiO$_2$/graphene composites forphotodegradation of methyl orange. *Nano Res.,* **2011**, *4*, 274-283.
       [http://dx.doi.org/10.1007/s12274-010-0079-4]

[107] Zhang, Y.; Tang, Z-R.; Fu, X.; Xu, Y-J. TiO$_2$-graphene nanocomposites for gas-phase photocatalytic degradation of volatile aromatic pollutant: is TiO$_2$-graphene truly different from other TiO$_2$-carbon composite materials? *ACS Nano,* **2010**, *4*(12), 7303-7314.
       [http://dx.doi.org/10.1021/nn1024219] [PMID: 21117654]

[108] Morales Torres, S. PastranaMartínez, L.M., Figueiredo, J.L., Faria, J.L., Silva. *A.M.T. Environ. Sci. Pollut. R.,* **2012**, *19*, 3676-3687.

[109] Wang, X.; Zhi, L.; Müllen, K. Transparent, conductive graphene electrodes for dye-sensitized solar cells. *Nano Lett.,* **2008**, *8*(1), 323-327.
       [http://dx.doi.org/10.1021/nl072838r] [PMID: 18069877]

[110] Wang, Y.; Chen, Y.-X.; Barakat, T.; Wang, T.-M.; Krief, A.; Zeng, Y.-J. Synergistic effects of carbon doping and coating of TiO$_2$ with exceptional photocurrent enhancement for high performance H2 production from water splitting. *J. Energy Chem,* **2020**.
       [http://dx.doi.org/10.1016/j.jechem.2020.08.002]

[111] Bellamkonda, S.; Thangavel, N.; Hafeez, H.Y.; Neppolian, B.; Rao, R. G. Highly active and stable multi-walled carbon nanotubes-graphene-TiO 2 nanohybrid: An efficient non-noble metal photocatalyst for water splitting. *Catal. Today,* **2017**.
[http://dx.doi.org/10.1016/j.cattod.2017.10.023]

[112] Cheng, P.; Yang, Z.; Wang, H.; Cheng, W.; Chen, M.; Shangguan, W.; Ding, G. TiO_2–graphene nanocomposites for photocatalytic hydrogen production from splitting water. *Int. J. Hydrogen Energy,* **2012**, *37*(3), 2224-2230.
[http://dx.doi.org/10.1016/j.ijhydene.2011.11.004]

[113] Min, S.; Wang, F.; Lu, G. Graphene-induced spatial charge separation for selective water splitting over TiO 2 photocatalyst. *Catal. Commun.,* **2016**, *80*, 28-32.
[http://dx.doi.org/10.1016/j.catcom.2016.03.015]

[114] Mizuno, R.; Tsuchihashi, R.; Furukawa, S.; Takano, H.; Takase, M.; Kuga, Y.; Yamanaka, S. Preparation of concentrated multilayer graphene dispersions and TiO_2-graphene composites for enhanced hydrogen production. *Diamond Related Materials,* **2019**, *107516*
[http://dx.doi.org/10.1016/j.diamond.2019.107516]

[115] Zhang, X.Y.; Sun, Y.J. A green and facile synthesis of TiO_2/ graphene nanocomposites and their photocatalytic activity for hydrogen evolution. *Int. J. Hydrogen Energy,* **2012**, *37*, 811-815.
[http://dx.doi.org/10.1016/j.ijhydene.2011.04.053]

[116] Fan, W.Q.; Lai, Q.H.; Zhang, Q.H.; Wang, Y. Nanocomposites of TiO_2 and reduced graphene oxide as efficient photocatalysts for hydrogen evolution. *J. Phys. Chem. C,* **2011**, *115*, 10694-106701.
[http://dx.doi.org/10.1021/jp2008804]

[117] Zhang, X.Y.; Li, H.P.; Cui, X.L.; Lin, Y. Graphene/TiO_2 nanocomposites: synthesized, characterization and application in hydrogen evolution from water photocatalytic splitting. *J. Mater. Chem.,* **2010**, *20*, 2801-2806.
[http://dx.doi.org/10.1039/b917240h]

[118] Shen, J.; Yan, B.; Shi, M.; Ma, H.; Li, N.; Ye, M. One step hydrothermal synthesis of TiO_2-reduced graphene oxide sheets. *J. Mater. Chem.,* **2011**, *21*, 3415-3421.
[http://dx.doi.org/10.1039/c0jm03542d]

[119] Akhavan, O. Graphene nanomesh by ZnO nanorod photocatalysts. *ACS Nano,* **2010**, *4*(7), 4174-4180.
[http://dx.doi.org/10.1021/nn1007429] [PMID: 20550104]

[120] Williams, G.; Seger, B.; Kamat, P.V. TiO_2-graphene nanocomposites. UV-assisted photocatalytic reduction of graphene oxide. *ACS Nano,* **2008**, *2*(7), 1487-1491.
[http://dx.doi.org/10.1021/nn800251f] [PMID: 19206319]

[121] Zhou, Y.; Bao, Q.; Tang, L.A.L.; Zhong, Y.; Loh, K.P. Hydrothermal dehydration for the "green" reduction of exfoliated graphene oxide to graphene and demonstration of tunable optical limiting properties. *Chem. Mater.,* **2009**, *21*, 2950-2956.
[http://dx.doi.org/10.1021/cm9006603]

[122] Wang, P.; Zhan, S.; Xia, Y. The Fundamental Role and Mechanism of Reduced Graphene Oxide in rGO/Pt-TiO_2 Nanocomposite for High-performance Photocatalytic Water Splitting *Appl.Catal.B: Environ,* http://dx.doi.org/doi:10.1016/j.apcatb

[123] Khalid, N.R.; Hong, Z.; Ahmed, E.; Zhang, Y.; Chan, H.; Ahmad, M. Synergistic effects of Fe and graphene on photocatalytic activity enhancement of TiO_2 under visible light. *Appl. Surf. Sci.,* **2012**, *258*(15), 5827-5834.
[http://dx.doi.org/10.1016/j.apsusc.2012.02.110]

[124] Xiang, Q.; Yu, J.; Jaroniec, M. Synergetic effect of MoS2 and graphene as cocatalysts for enhanced photocatalytic H_2 production activity of TiO_2 nanoparticles. *J. Am. Chem. Soc.,* **2012**, *134*(15), 6575-6578.
[http://dx.doi.org/10.1021/ja302846n] [PMID: 22458309]

[125] Iwase, A.; Ng, Y.H.; Ishiguro, Y.; Kudo, A.; Amal, R. Reduced graphene oxide as a solid-state electron mediator in Z-scheme photocatalytic water splitting under visible light. *J. Am. Chem. Soc.,* **2011**, *133*(29), 11054-11057.
[http://dx.doi.org/10.1021/ja203296z] [PMID: 21711031]

[126] Ling, T.; Song, J-G.; Chen, X-Y.; Yang, J.; Qiao, S-Z.; Du, X-W. Comparison of ZnO and TiO$_2$ nanowires for photoan ode of dye-sensitized solar cells. *J. Alloys Compd.,* **2013**, *546*, 307-313.
[http://dx.doi.org/10.1016/j.jallcom.2012.08.030]

CHAPTER 10

Catalytic Applications of Graphene Nanomaterials in Organic Transformation

Somit Kumar Singh[1,*]

[1] *Department of Chemistry, College of Natural and Mathematical Sciences, The University of Dodoma, PO Box 259, Tanzania*

Abstract: This chapter is based on the catalytic application of various graphene nanomaterials in organic synthesis and group transformations. Graphene oxide is a good heterogeneous catalyst due to the presence of oxygen containing functional groups. Polyphenolic compounds from plant extracts act as an efficient, reducing agent for bio-reduction of graphene oxide. The reduction abilities of such phytochemicals have been reported in the synthesis and stabilization of various nanoparticles *viz.* Ag, Au, Fe and Pd. Furthermore, the chapter describes the catalytic applications of graphene, graphene oxide, reduced graphene oxide nanosheets and metal based reduced graphene oxide nanoparticles as efficient carbo-catalysts for valuable organic transformations including oxidation, carbon–carbon cross-coupling, esterification, aza-Michael addition, Knoevenagel condensation, Friedel-Crafts addition, acetylation, thiocyanization, Alcoholysis, oxidative coupling of amine, Suzuki-Miyaura coupling, Heck and Sonogashira coupling, Diels-Alder reaction, ring opening of epoxides, hydrogenation/reduction and a wide number of multicomponent reactions. The applications of graphene-based catalysts allow mild, selective and highly effective transformations and synthesis in a facile, regenerable, recyclable and eco-friendly manner.

Keywords: Catalysis, Graphene, Nanomaterials, Organic transformation.

INTRODUCTION

Graphene is an allotrope of carbon consisting of a single layer of atoms arranged in a two-dimensional honeycomb lattice [1]. Each atom in a graphene sheet is connected to its three nearest neighbors by an σ-bond, and contributes one electron to a conduction band that extends over the whole sheet. These conduction bands make graphene a semimetal with unusual electronic properties that are best described by theories for massless relativistic particles [2]. Graphene conducts

* **Corresponding author Somit K Singh:** Department of Chemistry, College of Natural and Mathematical Sciences, The University of Dodoma, PO Box 259, Tanzania; Tel: +255754475477; Fax: +255 0262310005; E-mail: somitsingh@gmail.com

Manorama Singh, Vijai K. Rai and Ankita Rai (Eds)

heat and electricity very efficiently along its plane. The material strongly absorbs the light of all visible wavelengths [3, 4], which accounts for the black color of graphite; yet a single graphene sheet is nearly transparent because of its extreme thinness. The material is also about 100 times stronger than would be the strongest steel of the same thickness [5, 6]. High-quality graphene isolated and characterized in 2004 by A. Geim and K. Novoselov, they pulled graphene layers from graphite with a common adhesive tape in a process called either micromechanical cleavage or the Scotch tape technique [7]. Graphene is a zero-gap semiconductor, because its conduction and valence bands meet at the Dirac points and its hexagonal lattice can be regarded as two interleaving triangular lattices. This perspective was successfully used to calculate the band structure for a single graphite layer using a tight-binding approximation [8]. Graphene's permittivity varies with frequency over a range from microwave to millimeter wave frequencies. This permittivity, combined with the ability to form conductors and insulators, means that theoretically, compact capacitors made of graphene could store large amounts of electrical energy [9]. Multi-Parametric Surface Plasmon Resonance (MPSPR) was used to characterize the thickness and refractive index of chemical-vapor-deposition (CVD)-grown graphene films. The thickness was determined as 3.7Å from a 0.5 mm area, which agrees with 3.35Å reported for layer-to-layer carbon atom distance of graphite crystals [10]. The method can be further used for real-time label-free interactions of graphene with organic and inorganic substances. Furthermore, the existence of unidirectional surface plasmons in the nonreciprocal graphene-based gyrotropic interfaces has been demonstrated theoretically. By efficiently controlling the chemical potential of graphene, the unidirectional working frequency can be continuously tunable from terahertz to near-infrared and even visible [11]. Graphene's band gap can be tuned from 0 to 0.25 eV (about 5 micrometer wavelength) by applying a voltage to a dual-gate bilayer graphene field-effect transistor (FET) at room temperature [12].

GENERAL INFORMATION OF GRAPHENE BASED NANOMATERIALS

Graphene, a two-dimensional single-layer carbon sheet with hexagonal packed lattice structure, serves as a potential building block for fullerenes, nanotubes and graphite having zero-, one-, two- and three-dimensional structures, respectively [2]. High-quality large-area graphene sheets can be obtained by epitaxial growth of single-crystal silicon carbide (SiC) *via* ultrahigh vacuum annealing [13]. Nevertheless, the method has certain disadvantages and found unsuitable for large-scale production. In this context, the most accepted model dealing with GO structure is the one by Lerf, Klinowski and coworkers [14] (Fig. **1**), where it is proposed that the heavily oxygenated GO consists of hydroxyl and epoxide groups on the basal planes, while carboxyl at the edges.

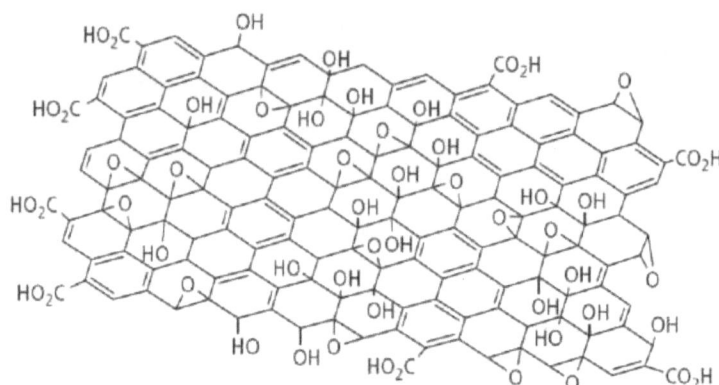

Fig. (1). Structural model of graphene oxide. Image adapted from [14].

In recent times, graphene generated from graphitic and non-graphitic materials, the most developed method to generate single-layered graphene from graphitic materials in excellent yields are; Mechanical exfoliation and cleavage [15], Liquid phase mechanical exfoliation [16], Graphite intercalation compounds [17], Pyrolysis, Electrochemical exfoliation [18] and Sonication [19]. Graphene has been developed by non-graphitic materials by chemical vapor deposition (CVD) [20], epitaxial growth from silicon carbide [21] and un-zipping CNTs [22].

Graphene oxide (GO) synthesized by the addition of $KClO_3$ to the slurry of graphite in fuming HNO_3 [23]. Afterward, it was enhanced by the addition of chlorate to the reaction mixture along with concentrated sulfuric acid and fuming nitric acid very slowly [24]. In 1958, Hummers reported the method most commonly used today; the graphite is oxidized by treatment with $KMnO_4$ and $NaNO_3$ in concentrated H_2SO_4 [25].

Reduced graphene oxide (rGO) obtained from the chemical/thermal reduction of GO has been reported as carbocatalyst for a variety of chemical reactions. The chemical reduction of graphene oxide was developed by using hydrazine, dimethylhydrazine, hydroquinone and $NaBH_4$ as the reducing agent. Investigation on replacing the current reduction methods by some biocompatible, chemical and impurity-free rGO including; Flash photo reductions [26], Hydrothermal dehydration [27], Solvothermal reduction [28], Electrochemical approach [29], Microwave-assisted reductions [30] and Light and radiation-induced reductions [31], have been reported. A large number of natural antioxidants such as vitamins, amino acids, proteins and organic acids readily available and utilized to reduce, GO successfully. Various parts of plant leaf extracts such as tea leaf extract [32], *Ginkgo biloba* leaf extract [33], cherry leaf extract [34], *Platanus orientalis* leaf extract [35], *Colocasia esculenta* leaf extract [36], *Spinacia oleracea* leaf extract

[37], aloe vera leaf extract [38] and eucalyptus leaf extract [39] have been applied for the green reduction of graphene oxide into reduced graphene oxide.

CATALYTIC APPLICATIONS OF GRAPHENE BASED NANOMATERIALS IN ORGANIC TRANSFORMATION

Some important work dedicated to exploring graphene-based nanomaterials (graphene oxide and reduced graphene oxide nanosheets) as carbo-catalysts, used in many catalytic based organic synthesis, such as catalytic oxidations, reduction, additions, group transfer reactions, multicomponent reactions, *etc.* (Fig. **2**).

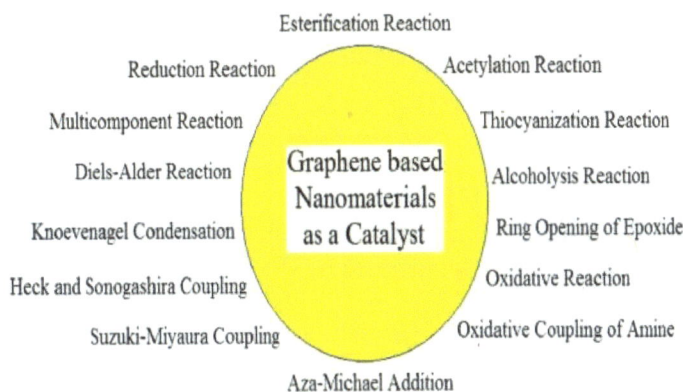

Esterification Reaction

Reduction Reaction   Acetylation Reaction

Multicomponent Reaction   Thiocyanization Reaction

Diels-Alder Reaction   **Graphene based**   Alcoholysis Reaction
                              Nanomaterials
Knoevenagel Condensation   **as a Catalyst**   Ring Opening of Epoxide

Heck and Sonogashira Coupling   Oxidative Reaction

Suzuki-Miyaura Coupling   Oxidative Coupling of Amine

Aza-Michael Addition

Fig. (2). Schematic Illustration of Typical Organic Reactions *versus* Graphene-based Nanomaterials Catalysis.

Esterification Reaction

Various carboxylic acids and alcohols undergo esterification in the presence of GO as a catalyst under reflux to obtain corresponding ester in high yields [40] (Scheme **1**).

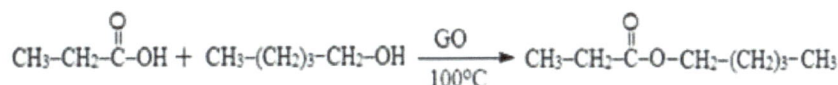

$$CH_3-CH_2-\overset{O}{\overset{\|}{C}}-OH + CH_3-(CH_2)_3-CH_2-OH \xrightarrow[100°C]{GO} CH_3-CH_2-\overset{O}{\overset{\|}{C}}-O-CH_2-(CH_2)_3-CH_3$$

Scheme (1). Esterification of Propionic acid and Pentanol in to Corresponding Ester.

Acetylation Reaction

Acetylation of aldehydes in methanol (Scheme **2**) at room temperature in the presence of exfoliated graphene oxide as heterogeneous catalyst has been well reported by H. Garcia and co-workers [41].

$$CH_3-CHO + CH_3OH \xrightarrow[RT]{\text{Exfoliated GO}} CH_2\!\!\begin{array}{c} O-CH_3 \\ O-CH_3 \end{array}$$

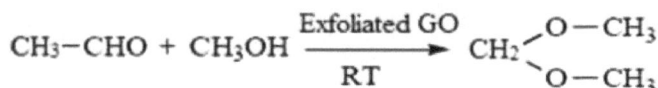

Scheme (2). Acetylation of Acetaldehyde.

Thioacetalization

Synthesis of open chain, cyclic and unsymmetrical dithioacetals from aryl, hetero-aryl/aliphatic aldehydes catalyzed by GO (Scheme **3**) [42].

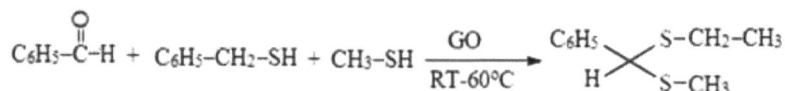

$$C_6H_5-\overset{\overset{\displaystyle O}{\|}}{C}-H + C_6H_5-CH_2-SH + CH_3-SH \xrightarrow[RT-60^\circ C]{GO} \begin{array}{c} C_6H_5 \\ H \end{array}\!\!\!\!\times\!\!\!\!\begin{array}{c} S-CH_2-CH_3 \\ S-CH_3 \end{array}$$

Scheme (3). Synthesis of Dithioacetal.

Dehydrative Esterification

Synthesis of dibenzyl ether (DE) from benzyl alcohol by solvent-free catalytic dehydrative esterification in the presence of graphene oxide as carbocatalyst has been reported by Wang *et al.* (Scheme **4**) [43].

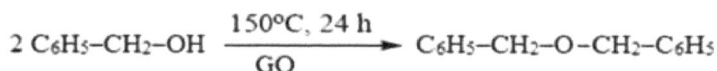

$$2\ C_6H_5-CH_2-OH \xrightarrow[GO]{150^\circ C,\ 24\ h} C_6H_5-CH_2-O-CH_2-C_6H_5$$

Scheme (4). Dehydrative Esterification of Benzyl Alcohol.

Trans Acetalization

1,2 and 1,3-diols are readily protected as cyclic acetals and ketals through graphene-catalyzed trans acetalization process. The graphene-catalyzed trans acetalization was performed under solvent-free conditions (Scheme **5**) [44].

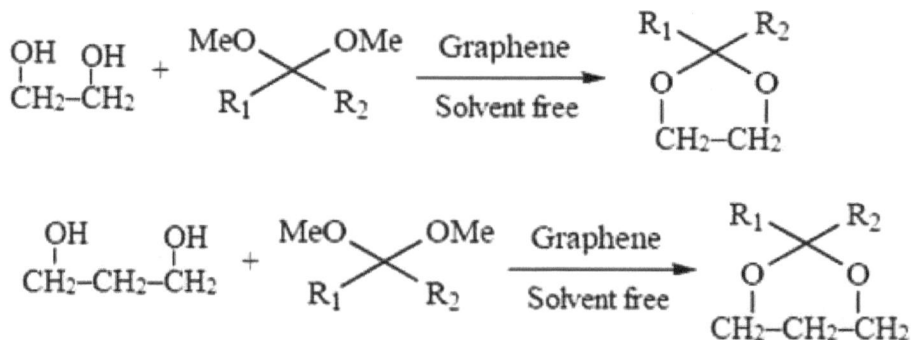

Scheme (5). Transacetalization of 1,2- and 1,3-Diol.

Thiocyanization

Arenes and carbonyl compounds having α-hydrogen undergo regioselective thiocynation with potassium thiocyanide and H_2O_2 in the presence of graphene oxide as a carbocatalyst (Scheme **6**) [45].

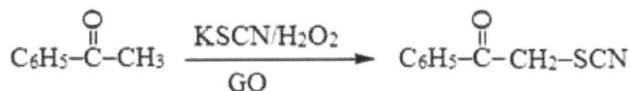

$$C_6H_5-\overset{\overset{O}{\|}}{C}-CH_3 \xrightarrow[GO]{KSCN/H_2O_2} C_6H_5-\overset{\overset{O}{\|}}{C}-CH_2-SCN$$

Scheme (6). Thiocyanation of Carbonyl Compound.

Alcoholysis Reaction

Furfuryl alcohol adsorbed on GO through surface-active SO_3H groups and undergo alcoholysis (Scheme **7**) [46].

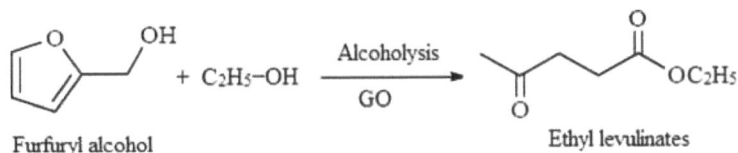

Furfuryl alcohol   Ethyl levulinates

Scheme (7). Alcoholysis of Furfuryl Alcohol.

Transamidation of Aliphatic Amides

Tertiary amides and aromatic amines interact with each other and converted into secondary amide by transamidation reaction in the presence of graphene oxide as an alternative heterogeneous catalyst (Scheme **8**) [47].

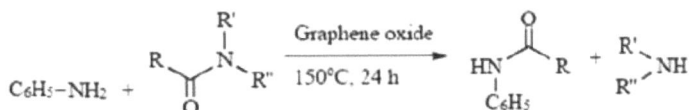

Scheme (8). Transamidation of Tertiary Amide in to Secondary Amide.

Dehydrogenation-Hydrothiolation of Secondary Aryl Alcohols

Various unsymmetrical thioethers synthesized by dehydrogenation-hydrothiolation reaction of secondary aryl alcohols and thiols in the presence of graphene oxide as a carbocatalyst (Scheme **9**) [48].

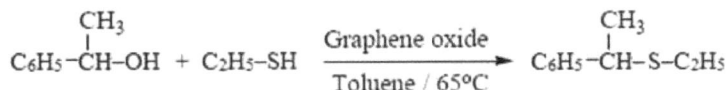

$$C_6H_5-\overset{\overset{CH_3}{|}}{C}H-OH + C_2H_5-SH \xrightarrow[\text{Toluene} / 65°C]{\text{Graphene oxide}} C_6H_5-\overset{\overset{CH_3}{|}}{C}H-S-C_2H_5$$

Scheme (9). Dehydrogenation- hydrogenation of Secondary Aryl Alcohol.

Ring-opening of Epoxides

Room temperature ring opening of epoxide (Scheme **10**) by using graphene oxide as a reusable acid catalyst has been reported by A. Dhakshinamoorthy and co-workers [49].

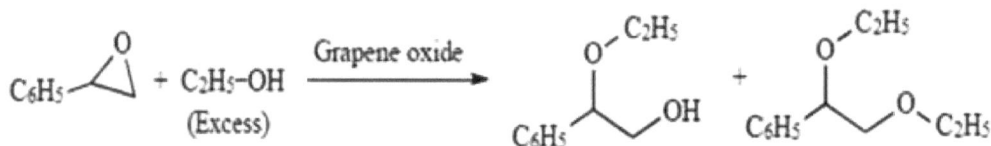

Scheme (10). Ring Opening Epoxide.

Oxidation Reactions

Graphene oxide and reduced graphene oxide contain many functional groups having oxygen, which can easily adsorb the substrates over their surface through hydrogen bonding; therefore GO has great efficiency for aerobic oxidation.

Aerobic Oxidation of Cyclohexane

Aerobic oxidation of cyclohexane (Scheme **11**) in the presence of mesoporous graphene (sp2 hybridized carbon) and diamond (sp3 hybridized carbon) as a carbocatalyst have been reported [50]. A better result was found with sp2 hybridized carbons.

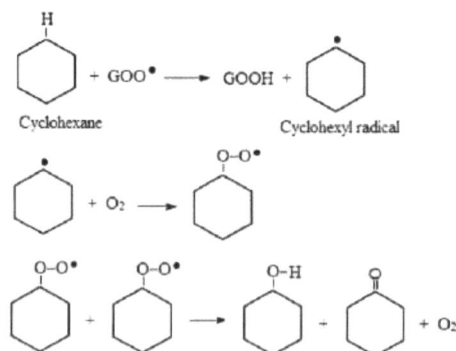

Scheme (11). Aerobic Oxidation of Cyclohexene.

C-H Oxidation

Oxidation of olefins, diarylmethanes and methyl benzenes into carbonyl compounds (Scheme **12**) in the presence of exploited graphene oxide as a catalyst has been reported [51].

$$C_6H_5-CH=CH-C_6H_5 \xrightarrow[\text{CHCl}_3]{\text{Graphene oxide}} C_6H_5-\overset{O}{\underset{||}{C}}-\overset{O}{\underset{||}{C}}-C_6H_5$$

$$C_6H_5-CH_2-C_6H_5 \xrightarrow{\text{Graphene oxide}} C_6H_5-\overset{O}{\underset{||}{C}}-C_6H_5$$

$$C_6H_5-CH_3 \xrightarrow{\text{Graphene oxide}} C_6H_5-\overset{O}{\underset{||}{C}}-H$$

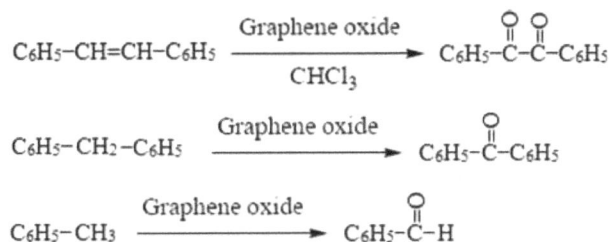

Scheme (12). Oxidation of Diarylolefin, Diarylmethane and Arylmethane.

Oxidation of Thiols

Oxidation of thiols to disulfides (Scheme **13**) in the presence of graphite oxide as low cost heterogeneous carbocatalyst has been reported [52]. Over-oxidation of the substrate was not observed in this process.

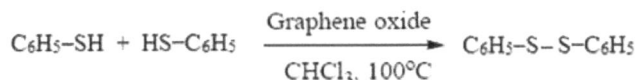

$$C_6H_5-SH + HS-C_6H_5 \xrightarrow[\text{CHCl}_3, 100°C]{\text{Graphene oxide}} C_6H_5-S-S-C_6H_5$$

Scheme (13). Oxidation of Thiols in to Disulphide.

Oxidative Aromatization of 1,4-dihydropyridines into Pyridine Derivatives

The oxidative aromatization of symmetrical and unsymmetrical 1,4-dihydropyridines [53] into corresponding pyridine derivatives was obtained in the presence of graphene oxide as an oxidizing catalyst (Scheme **14**).

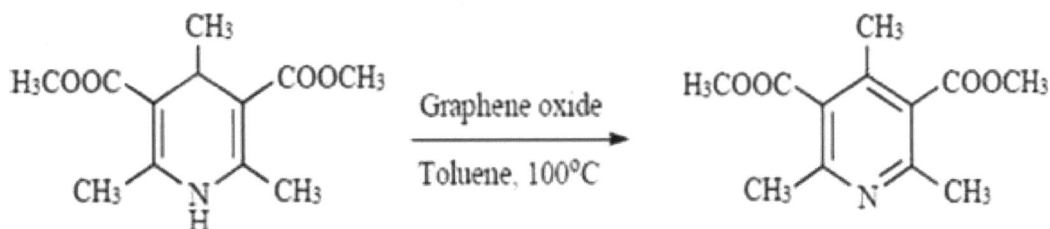

Scheme (14). Oxidative Aeromatization of 1,4-Dihydropyridine.

Oxidation of Alcohols

Oxidation of aromatic, heterocyclic, and aliphatic alcohols into aldehydes and ketones (Scheme **15**) in the presence of graphite oxide has been reported by Aghayan and co-workers [54]. GO has several advantages, such as low-cost, high stability and easy recovery from the reaction mixture over the other reagents applied for the same transformation.

$$C_6H_5-CH_2-OH \xrightarrow{\text{Graphene Oxide}} C_6H_5-CHO$$

$$\overset{\displaystyle OH}{\underset{\displaystyle }{C_2H_5-CH-CH_3}} \xrightarrow{\text{Graphene Oxide}} C_2H_5-\overset{\displaystyle O}{\overset{\displaystyle \|}{C}}-CH_3$$

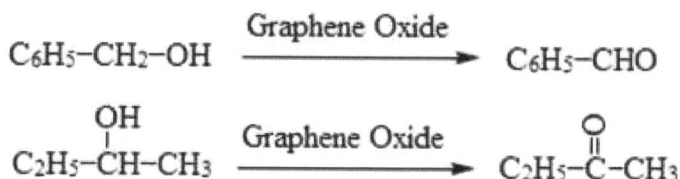

Scheme (15). Oxidation of Aromatic and Aliphatic Alcohol.

Oxidation of Benzene in to Phenol

Direct oxidation of benzene in to phenol (Scheme **16**) in the presence of H_2O_2 as an oxidant and chemically converted graphene (CCG) as a carbocatalyst [55]. Catalyst was highly stable and easily recovered from the reaction mixture.

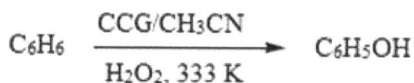

$$C_6H_6 \xrightarrow[H_2O_2,\ 333\ K]{CCG/CH_3CN} C_6H_5OH$$

Scheme (16). Direct Oxidation of Benzene.

Transformation of Hydroquinone into Benzoquinone

The oxidative transformation of 1,4-hydroquinone into 1,4-benzoquinone (Scheme **17**) dramatically increases under aerobic condition in the presence of reduced graphene oxide as a catalyst in an aqueous medium [56].

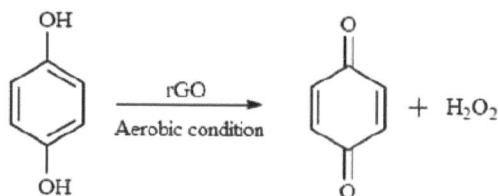

Scheme (17). Oxidation of Hydroquinone.

Oxidative Dehydrogenation of Hydrazo Compounds

Aerobic oxidative dehydrogenation of aromatic hydrazo compounds to their corresponding hydrazo compounds (Scheme **18**) in the presence of reduced graphene oxide as an efficient catalyst under mild conditions has been reported [57].

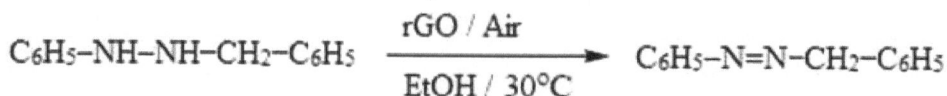

$$C_6H_5-NH-NH-CH_2-C_6H_5 \xrightarrow[EtOH\ /\ 30^\circ C]{rGO\ /\ Air} C_6H_5-N=N-CH_2-C_6H_5$$

Scheme (18). Oxidative Dehydrogenation of Hydrazo Compound.

Oxidation of Glutaraldehyde

Selective oxidation of glutaraldehyde into glutaric acid (Scheme **19**) with aqueous H_2O_2 in the presence of graphite oxide has been reported by Dai *et al.* [58].

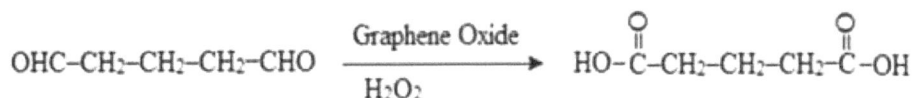

$$OHC-CH_2-CH_2-CH_2-CHO \xrightarrow[H_2O_2]{\text{Graphene Oxide}} HO-\overset{O}{\overset{\|}{C}}-CH_2-CH_2-CH_2-\overset{O}{\overset{\|}{C}}-OH$$

Scheme (19). Oxidation of Glutaraldehyde in to Glutaric Acid.

Oxidative Coupling of Amines

Oxidative coupling of amines into corresponding imines (Scheme **20**) by molecular oxygen as the terminal oxidant and graphite oxide as catalyst has been reported [59].

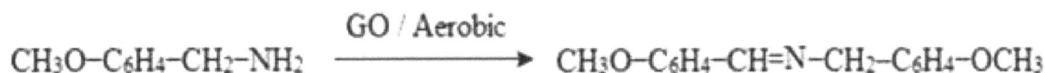

$$CH_3O-C_6H_4-CH_2-NH_2 \xrightarrow{\text{GO / Aerobic}} CH_3O-C_6H_4-CH=N-CH_2-C_6H_4-OCH_3$$

Scheme (20). Synthesis of Imine by Oxidative Coupling of Benzyl Amine.

Graphene Nanomaterials Catalyzed Reduction Reactions

Hydrogenation of Nitrobenzene

Reduction of nitrobenzene into aniline at room temperature (Scheme **21**) in the presence of reduced graphene oxide as an efficient catalyst has been studied by Gao and his co-workers [60]. Edges of rGO act as active sites to facilitate the activation of reactant molecules.

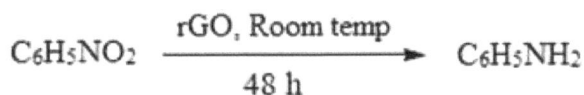

$$C_6H_5NO_2 \xrightarrow[48 \text{ h}]{\text{rGO, Room temp}} C_6H_5NH_2$$

Scheme (21). Synthesis of Aniline from Nitrobenzene.

Hydrogenation of Ethylene

Perhun *et al.* [61] investigated the catalytic properties of graphene materials in the hydrogenation of alkynes and ethylene to produce methane and ethane, respectively (Scheme **22**).

$$C_2H_4 \; + \; \boxed{} \; \longrightarrow \; \overset{\textstyle C_2H_4}{\boxed{}}$$

$$H_2 \; + \; \boxed{} \; \longrightarrow \; \overset{\textstyle H_2}{\boxed{}}$$

$$\overset{\textstyle C_2H_4}{\boxed{}} \; + \; \overset{\textstyle H_2}{\boxed{}} \; \longrightarrow \; C_2H_6 \; + \; 2 \; \boxed{}$$

$$\boxed{} \; \Longrightarrow \text{Active sites}$$

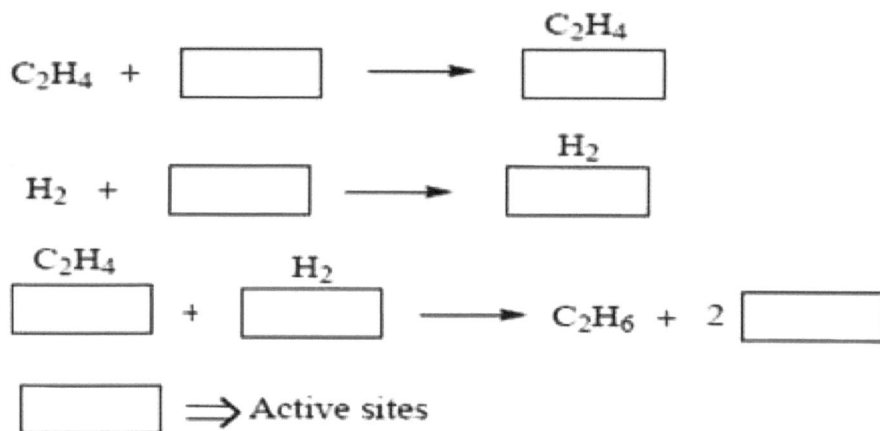

Scheme (22). Hydrogenation of Ethylene in to Ethane.

Hydrogenation of Nitrogen Heterocycles

Various nitrogen containing heterocyclic compounds, such as quinoline, 3,4-dihydroisoquinoline, quinazoline, and indole derivatives, undergo hydrogenation (Scheme **23**) significantly in the presence of graphene oxide as a cost-effective, metal-free carbocatalyst. Oxygen containing functional groups and large π-conjugated system of catalyst facilitate this transformation [62].

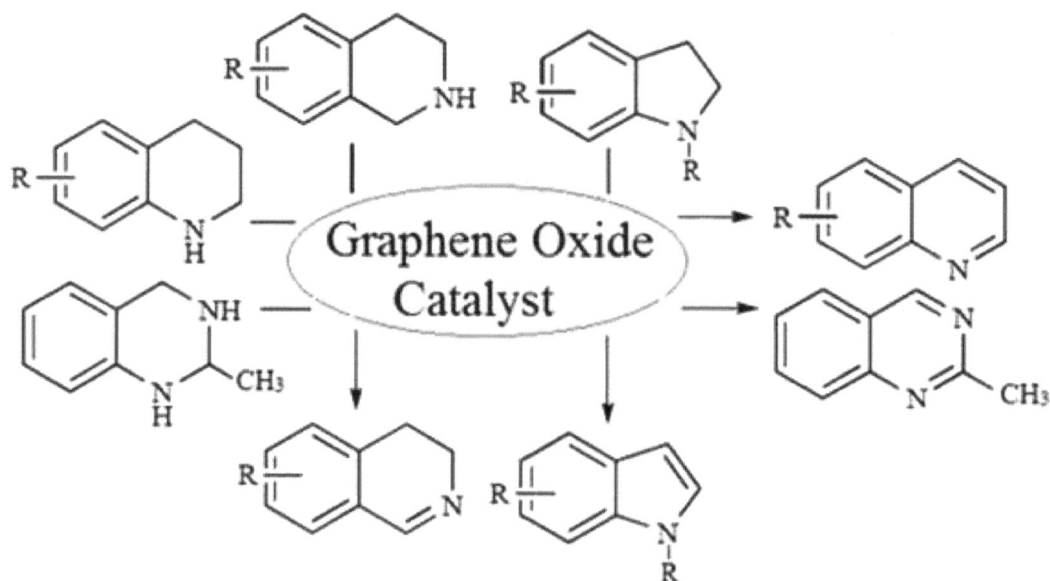

Scheme (23). Hydrogenation of Nitrogen Containing Heterocyclic Compounds.

Reduction of α, β-unsaturated Aldehydes

Graphene oxide catalyzed reduction of α, β-unsaturated aldehydes into allylic alcohols (Scheme **24**) in an aqueous medium has been reported by Rai *et al.* [63] with excellent yields of pure products (92%) in addition to regioselective products.

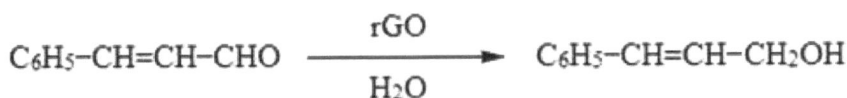

$$C_6H_5-CH=CH-CHO \xrightarrow[\text{H}_2\text{O}]{\text{rGO}} C_6H_5-CH=CH-CH_2OH$$

Scheme (24). Synthesis of Allylic Alcohol from α,β-Unsaturated Aldehyde.

Reduction of Methylene Blue

Methylene blue, under acidic conditions in the presence of reduced graphene as a catalyst, reduced into the colorless hydrogenated leucomethylene blue (Scheme **25**). Wu *et al.* [64] synthesized Fe_3O_4-Pt/RGO composite, which exhibited higher catalytic activity for the reduction of methylene blue. The recycling of the composite could be achieved by simply applying an external magnetic field.

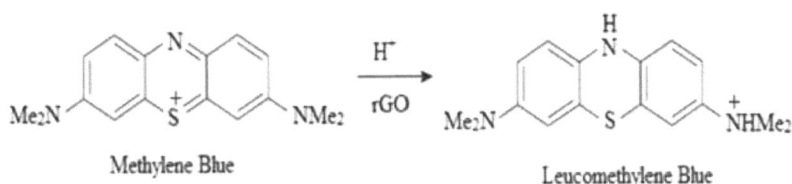

Scheme (25). Reduction of Methylene Blue in to Leucomethylene Blue.

Graphene Nanomaterials Catalyzed Coupling Reactions

Aza-Michael Addition of Amines

Synthesis of amino-substituted compounds *via* Aza-Michael addition of amines with vinyl derivatives (Scheme **26**) in the presence of graphene oxide as a highly efficient and cost-effective organocatalyst has been reported by Khatri *et al.* [65]. This reaction was the first ever reported coupling type reaction using graphene oxide as a reusable catalyst.

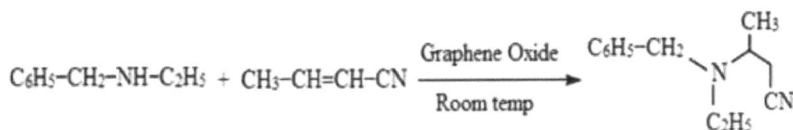

Scheme (26). Synthesis of Amine Substituted Compound.

Oxidative Dehydrogenative C-N Coupling

Dehydrogenative coupling of activated aldehydes and amines into α-ketoamides in the presence of graphene oxide as an acidic heterogeneous catalyst as well as oxidant has been investigated [66]. In this coupling, GO facilitates the formation of hemiaminal intermediate followed by oxidation leads to the formation of the target product. Primary and secondary amines are suitable for this reaction (Scheme **27**).

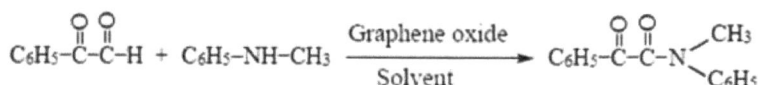

Scheme (27). Synthesis of α-Ketoamide from Activated Aldehyde and Amine.

Formation of Biaryl Compounds

Direct coupling of C-H bond of aryl iodide and benzene in the presence of graphene oxide as a green catalyst to form biaryl compound (Scheme **28**) has been reported by Wang *et al.* [67]. Such a type of reaction is also known as arylation of benzene.

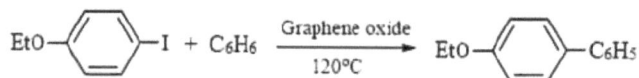

Scheme (28). Synthesis of Biaryl Compound.

Direct CH-CH Cross-coupling

Direct CH-CH cross-coupling reaction of xanthane or thioxanthane with arenes (Scheme **29**) in the presence of graphene oxide catalyst has been reported by Wu *et al.* [68].

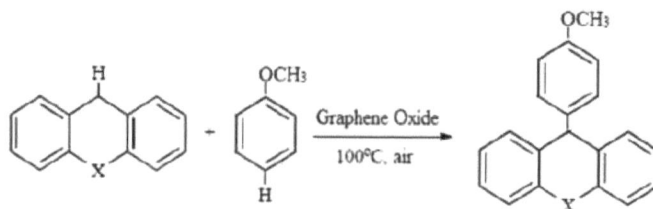

Scheme (29). Direct Cross-Coupling Reaction.

Suzuki-Miyaura Coupling

Aryl bromides interact with boric acids in the presence of Pd^{+2}-graphite oxide catalysts in aqueous solution to form corresponding coupling products in

81–100% yields (Scheme **30**) have been reported by Scheuermann *et al.* [69]. The catalysts were reusable, however, with some loss in the activity.

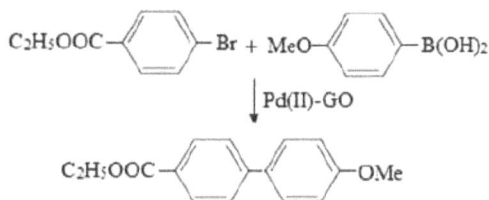

Scheme (30). Pd(II)-GO Mediated Suzuki-Miyaura Coupling.

Heck and Sonogashira Coupling

Coupling between butyl acrylate and aryl bromide in the presence of Pd (II)-graphite oxide as an organometallic heterogeneous catalyst to produce corresponding coupling products are Mizoroki-Heck and Sonogashira coupling reaction (Scheme **31**) [70].

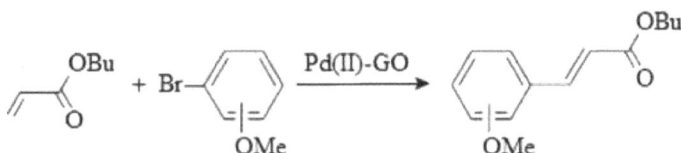

Scheme (31). Pd(II)-GO Mediated Mizoroki-Heck Coupling.

Graphene Nanomaterials Catalyzed Multicomponent Reactions

Friedel-Crafts Addition of Indoles

Graphene oxide catalyzed Friedel-Crafts addition of indoles with electron-rich and electron-deficient α, β-unsaturated ketones (Scheme **32**) has been reported [71] at an ambient condition of H_2O/THF solvent mixture.

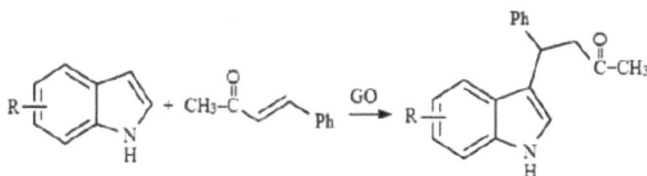

Scheme (32). Friedel-Crafts Addition of Indole and α,β-Unsaturated Ketone.

Direct Friedel-Crafts Alkylation Reactions

Graphene materials are applied for direct C-C bond formation by utilizing their polar functional groups. Alkylation of arenes with styrenes and alcohols in the presence of graphene materials as the catalyst for synthesizing valuable diarylalkane products (Scheme **33**) has been reported by Szostak *et al.* [72].

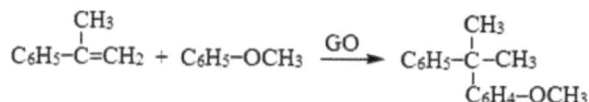

$$C_6H_5-\overset{\overset{\displaystyle CH_3}{|}}{C}=CH_2 + C_6H_5-OCH_3 \xrightarrow{\text{GO}} C_6H_5-\overset{\overset{\displaystyle CH_3}{|}}{\underset{\underset{\displaystyle C_6H_4-OCH_3}{|}}{C}}-CH_3$$

Scheme (33). Direct Friedel-Crafts Alkylation.

Regioselective Ring Opening of Aromatic Epoxide

Regio-and enantio-selective ring-opening reaction of aromatic epoxide by indole addition (Scheme **34**) in the presence of graphene oxide as a catalyst under solvent-free condition has been investigated by M. R. Acocella and his coworkers [73]. The reaction follows SN^2 mechanism, which ensures a high level of enantioselectivity.

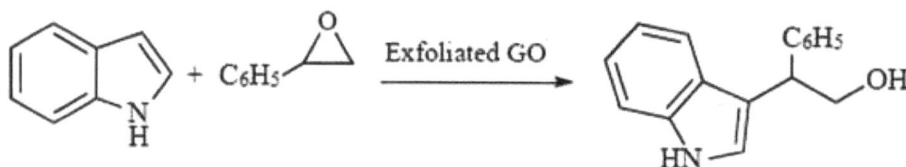

Scheme (34). Ring-Opening Reaction of Aromatic Epoxide by Indole Addition.

Knoevenagel Condensation

Knoevenagel condensation is a reaction between compounds containing active methylene group and aldchydes (Scheme **35**) in the presence of reduced graphene oxide as an active organocatalyst with yields up to 99% [74].

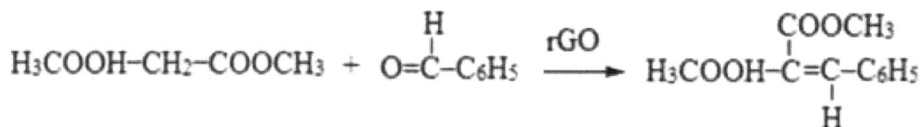

$$H_3COOH-CH_2-COOCH_3 + O=\overset{\overset{\displaystyle H}{|}}{C}-C_6H_5 \xrightarrow{\text{rGO}} H_3COOH-\overset{\overset{\displaystyle COOCH_3}{|}}{\underset{\underset{\displaystyle H}{|}}{C}}=C-C_6H_5$$

Scheme (35). Condensation of Compounds Containing Active Methylene and Aldehydes.

Michael Adducts

Graphene oxide used as a phase transfer catalyst with bases in H_2O and CH_2Cl_2 to obtained Michael adducts and their derivatives (Scheme **36**) reported by Y. Kim *et al.* [75]. The catalyst can be recovered and used several times without any loss of their catalytic activities.

$$C_6H_5-CH=CH-NO_2 \ + \ C_2H_5-\overset{\overset{O}{\|}}{C}-CH_2-\overset{\overset{O}{\|}}{C}-C_2H_5 \ \xrightarrow{GO} \ C_2H_5-\overset{\overset{O}{\|}}{C}-\underset{\underset{C_6H_5-CH-CH_2-NO_2}{|}}{CH}-\overset{\overset{O}{\|}}{C}-C_2H_5$$

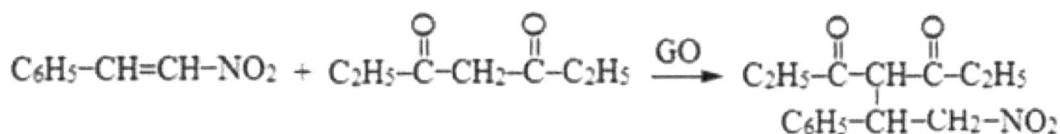

Scheme (36). Synthesis of Michael Adducts.

Synthesis of 5-Substituted 1H-Tetrazoles

Synthesis of 5-substituted 1H-tetrazoles by [2+3] cycloaddition reaction of nitriles and sodium azide (Scheme 37) in the presence of reduced graphene oxide as a catalyst with good yields has been reported [76].

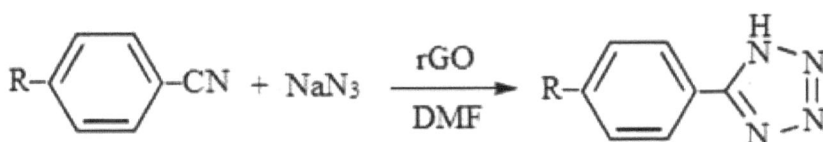

Scheme (37). Synthesis of 5-Substituted 1H-Tetrazoles.

Synthesis of Amides

Synthesis of amides using aromatic aldehydes and secondary amines (Scheme 38) at the surface of graphene oxide nanosheets has been investigated [77]. GO nanosheets were found to be an efficient, economic and metal free carbocatalyst because of the presence of much oxygen containing functional group.

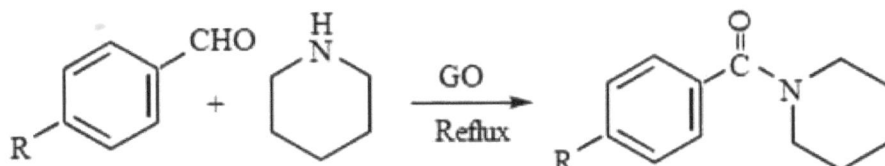

Scheme (38). Synthesis of Amide from Aldehyde and Secondary Amine.

Direct Oxidative Synthesis of Nitrones

Direct oxidative synthesis of various nitrones (Scheme 39) in good yields in the presence of graphene oxide and ozone as the oxidant as well as catalyst under mild reaction conditions was achieved [78]. This method revealed the condensation/oxidation of various aromatic and heteroaromatic aldehydes with aromatic and aliphatic amines.

$$C_6H_5CHO + C_2H_5NH_2 \xrightarrow{\text{GO/Ozone}} C_6H_5\text{—CH=}\overset{+}{N}\overset{\displaystyle C_2H_5}{\underset{\displaystyle O^-}{}}$$

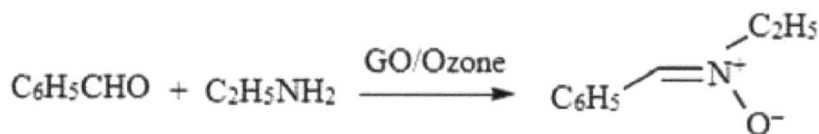

Scheme (39). Direct Oxidative Synthesis of Nitrones.

Synthesis of α-aminophosphonates

The Kabachnik-Fields reactions of aldehydes, amines and trimethylphosphite to produce α- aminophosphonates (Scheme **40**) in presence of graphene oxide as a recoverable catalyst at room temperature has been introduced by P. R. Nemade and co-workers [79].

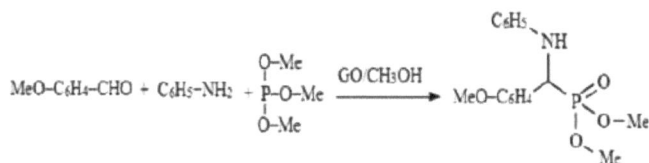

$$MeO\text{-}C_6H_4\text{-}CHO + C_6H_5\text{-}NH_2 + \overset{\displaystyle O\text{-}Me}{\underset{\displaystyle O\text{-}Me}{P\text{-}O\text{-}Me}} \xrightarrow{\text{GO/CH}_3\text{OH}} MeO\text{-}C_6H_4 \overset{\displaystyle C_6H_5\text{-}NH}{\underset{\displaystyle}{\text{—CH—}}}P\overset{\displaystyle O}{\underset{\displaystyle O\text{-}Me}{\text{—}O\text{-}Me}}$$

Scheme (40). Synthesis of α-Aminophosphonates by Aromatic Aldehyde and Amine.

Synthesis of 2-amino-3-cyanopyridines

Synthesis of 2-amino-3-cyanopyridine derivatives by aldehydes, amines and malononitrile (Scheme **41**) in the presence of graphene oxide as a catalyst has been reported [80]. Aqueous ammonium acetate was found to be the reaction medium.

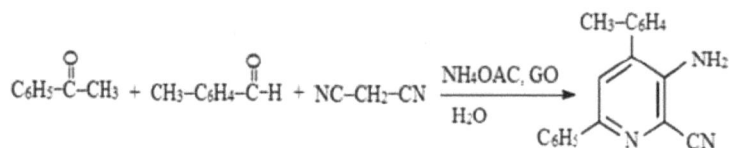

$$C_6H_5\text{-}\overset{\displaystyle O}{\overset{\|}{C}}\text{-}CH_3 + CH_3\text{-}C_6H_4\text{-}\overset{\displaystyle O}{\overset{\|}{C}}\text{-}H + NC\text{-}CH_2\text{-}CN \xrightarrow[\text{H}_2\text{O}]{\text{NH}_4\text{OAC, GO}}$$

Scheme (41). Synthesis of 2-Amino-3-Cyanopyridine by Aldehydes and Malononitrile.

Synthesis of Spiro Thiazolidinones Compounds

Synthesis of Spiro thiazolidinones compounds (Scheme **42**) can be obtained by dearomatization cyclizations of naphthol and Mannich bases with carbon disulfide in the presence of reduced graphene oxide under mild reaction conditions has been investigated [81].

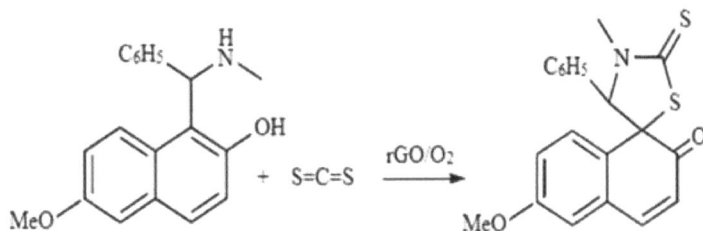

Scheme (42). Dearomatization Cyclizations of Naphthol and Mannich Bases with Carbon Disulphide.

Diels-Alder Reactions

Cycloaddition of 9-hydroxymethylanthracene and N-ethoxymaleimides (Scheme **43**) in the presence of graphene oxide as a catalyst in an aqueous medium under mild reaction condition has been studied by Dey *et al*. [82].

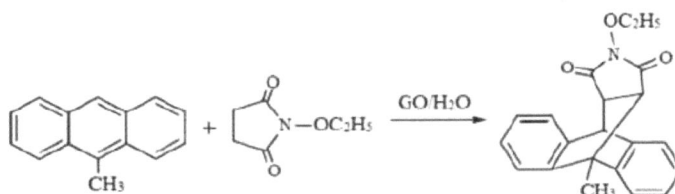

Scheme (43). Cycloaddition of 9-Hydroxymethylanthracene and N-Ethoxymaleimides.

Synthesis of Functionalized 1,4-benzothiazines

Green synthesis of highly functionalized heterocyclic compound 1,4-Benzothiazine from 2-aminothiophenol and 1,3-dicarbonyl compounds (Scheme **44**) in the presence of graphene oxide as a catalyst under solvent-free condition has been reported [83].

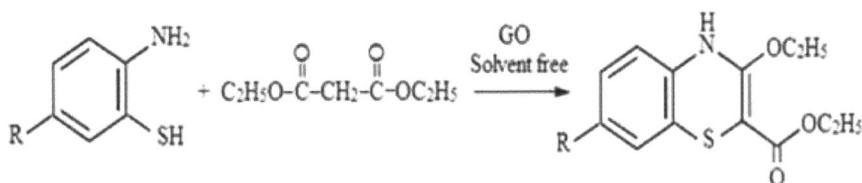

Scheme (44). Synthesis of 1,4-Benzothiazine from 2-Aminothiophenol and 1,3-Dicarbonyl Compounds.

Synthesis of 2,3-dihydroquinolinones and quinazolin-4(3H)-one

Anthranilamide (2-aminobenzamide) reacts with an aldehyde/ketone (Scheme **45**) in the presence of graphene oxide as a recyclable catalyst at room temperature in water to form 2,3-dihydroquinolinones and quinazolin-4(3H)-one [84]. This is the first application of GO for the selective cleavage of a C-C bond.

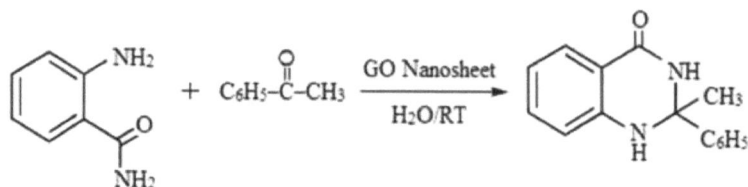

Scheme (45). Synthesis of 2,3-Dihydroquinolinones and Quinazoline-4(3H)-One.

Synthesis of Thiazolidine-4-one Derivatives

Green synthesis of bioactive 1,3-thiazolidine-4-one by using amines, aldehydes and mercapto acetic acid (Scheme **46**) in the presence of graphene oxide catalyst in aqueous medium has been reported by Rai *et al.* [85]. Product formation takes place *via* C-N and C-S bond formation and ring transformation.

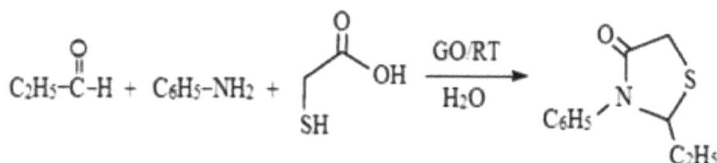

Scheme (46). Synthesis of 1,3-Thiazolidine-4-One by Amines, Aldehydes and Mercapto Acetic Acid.

CONCLUSION

In summary, recent progress in graphene-based nanomaterials has been presented. The advantages of the 2D structure of graphene and its properties have been discussed in this section. Structural defects and the presence of oxygen-containing functionalities on the catalytic activity of graphene and graphene oxide have been elaborated. Finally, description over the catalytic applications of graphene based nanomaterials in organic synthesis including oxidation, carbon–carbon cross-coupling, esterification, aza-Michael addition, Knoevenagel condensation, Friedel-Crafts addition, acetylation, thiocyanization, Alcoholysis, oxidative coupling of amine, Suzuki-Miyaura coupling, Heck and Sonogashira coupling, Diels-Alder reaction, ring opening of epoxides, hydrogenation/reduction and a wide number of multicomponent reactions have been discussed in this chapter.

CONSENT FOR PUBLICATION

Not applicable.

CONFLICT OF INTERESTS

The author declares no conflict of interest, financial or otherwise.

ACKNOWLEDGEMENTS

Author is thankful to the management of the University of Dodoma for providing all necessary facilities to complete this chapter.

REFERENCES

[1]   Peres, N.M.R.; Ribeiro, R.M. Focus on Graphene. *New J. Phys.,* **2009,** *11*(9), 095002.
       [http://dx.doi.org/10.1088/1367-2630/11/9/095002]

[2]   Geim, A.K.; Novoselov, K.S. The rise of graphene. *Nat. Mater.,* **2007,** *6*(3), 183-191.
       [http://dx.doi.org/10.1038/nmat1849] [PMID: 17330084]

[3]   Nair, R.R.; Blake, P.; Grigorenko, A.N.; Novoselov, K.S.; Booth, T.J.; Stauber, T.; Peres, N.M.R.;
       Geim, A.K. Fine structure constant defines visual transparency of graphene. *Science,* **2008,** *320*(5881),
       1308.
       [http://dx.doi.org/10.1126/science.1156965] [PMID: 18388259]

[4]   Zhu, S-E.; Yuan, S. Janssen, G. C. A. M. Optical transmittance of multilayer graphene. *EPL,* **2014,**
       108(1), 17007.
       [http://dx.doi.org/10.1209/0295-5075/108/17007]

[5]   Lee, C.; Wei, X.; Kysar, J.W.; Hone, J. Measurement of the elastic properties and intrinsic strength of
       monolayer graphene. *Science,* **2008,** *321*(5887), 385-388.
       [http://dx.doi.org/10.1126/science.1157996] [PMID: 18635798]

[6]   Cao, K.; Feng, S.; Han, Y.; Gao, L.; Hue Ly, T.; Xu, Z.; Lu, Y. Elastic straining of free-standing
       monolayer graphene. *Nat. Commun.,* **2020,** *11*(1), 284.
       [http://dx.doi.org/10.1038/s41467-019-14130-0] [PMID: 31941941]

[7]   Novoselov, K.S.; Geim, A.K.; Morozov, S.V.; Jiang, D.; Zhang, Y.; Dubonos, S.V.; Grigorieva, I.V.;
       Firsov, A.A. Electric field effect in atomically thin carbon films. *Science,* **2004,** *306*(5696), 666-669.
       [http://dx.doi.org/10.1126/science.1102896] [PMID: 15499015]

[8]   Cooper, D.R.; D'Anjou, B.; Ghattamaneni, N.; Harack, B. ilke, Michael; Horth, Alexandre; Majlis,
       Norberto; Massicotte, Mathieu; Vandsburger, Leron; Whiteway, Eric; Yu, Victor. "Experimental
       Review of Graphene. *Condens. Matter Phys.,* **2011,** 1-56.

[9]   Cismaru, Alina; Dragoman, Mircea; Dinescu, Adrian; Dragoman, Daniela; Stavrinidis, G;
       Konstantinidis, G Microwave and Millimeter wave Electrical Permittivity of Graphene Monolayer
       2013.

[10]   Jussila, H.; Yang, H.; Granqvist, N.; Sun, Z. Surface plasmon resonance for characterization of large-
       area atomic-layer graphene film. *Optica,* **2016,** *3*(2), 151-158.
       [http://dx.doi.org/10.1364/OPTICA.3.000151]

[11]   Lin, X.; Xu, Y.; Zhang, B.; Hao, R.; Chen, H.; Li, E. Unidirectional surface plasmons in nonreciprocal
       graphene. *New J. Phys.,* **2013,** *15*(11), 113003.
       [http://dx.doi.org/10.1088/1367-2630/15/11/113003]

[12]   Zhang, Y. Tang, Tsung-Ta; Girit, Caglar; Hao, Zhao; Martin, Michael C.; Zettl, Alex; Crommie,
       Michael F.; Shen, Y. Ron; Wang, Feng. "Direct observation of a widely tunable band gap in bilayer
       graphene. *Nature,* **2009,** *459*(7248), 820-823.
       [http://dx.doi.org/10.1038/nature08105] [PMID: 19516337]

[13]   Berger, C.; Song, Z.; Li, X.; Wu, X.; Brown, N.; Naud, C.; Mayou, D.; Li, T.; Hass, J.; Marchenkov,
       A.N.; Conrad, E.H.; First, P.N.; de Heer, W.A. Electronic confinement and coherence in patterned
       epitaxial graphene. *Science,* **2006,** *312*(5777), 1191-1196.
       [http://dx.doi.org/10.1126/science.1125925] [PMID: 16614173]

[14]   Lerf, A.; He, H.; Forster, M.; Klinowski, J. Structure of graphite oxide revisited. *J. Phys. Chem. B,*

1998, *102*, 4477-4482.
[http://dx.doi.org/10.1021/jp9731821]

[15]   Jayasena, B.; Subbiah, S. A novel mechanical cleavage method for synthesizing few-layer graphenes. *Nanoscale Res. Lett.,* **2011**, *6*(1), 95-102.
[http://dx.doi.org/10.1186/1556-276X-6-95] [PMID: 21711598]

[16]   Hernandez, Y.; Nicolosi, V.; Lotya, M.; Blighe, F.M.; Sun, Z.; De, S.; McGovern, I.T.; Holland, B.; Byrne, M.; Gun'Ko, Y.K.; Boland, J.J.; Niraj, P.; Duesberg, G.; Krishnamurthy, S.; Goodhue, R.; Hutchison, J.; Scardaci, V.; Ferrari, A.C.; Coleman, J.N. High-yield production of graphene by liquid-phase exfoliation of graphite. *Nat. Nanotechnol.,* **2008**, *3*(9), 563-568.
[http://dx.doi.org/10.1038/nnano.2008.215] [PMID: 18772919]

[17]   Dresselhaus, M.S.; Dresselhaus, G.; Dresselhaus, M.S.; Cançado, L.G.; Jorio, A.; Saito, R. Studying disorder in graphite-based systems by Raman spectroscopy. *Adv. Phys.,* **2002**, *51*(11), 1-186.
[http://dx.doi.org/10.1080/00018730110113644]

[18]   Parvez, K.; Li, R.; Puniredd, S.R.; Hernandez, Y.; Hinkel, F.; Wang, S.; Feng, X.; Müllen, K. Electrochemically exfoliated graphene as solution-processable, highly conductive electrodes for organic electronics. *ACS Nano,* **2013**, *7*(4), 3598-3606.
[http://dx.doi.org/10.1021/nn400576v] [PMID: 23531157]

[19]   Jiao, L.; Wang, X.; Diankov, G.; Wang, H.; Dai, H. Facile synthesis of high-quality graphene nanoribbons. *Nat. Nanotechnol.,* **2010**, *5*(5), 321-325.
[http://dx.doi.org/10.1038/nnano.2010.54] [PMID: 20364133]

[20]   Zheng, H.; Smith, R.K.; Jun, Y.W.; Kisielowski, C.; Dahmen, U.; Alivisatos, A.P.; Piner, R.; Velamakanni, A.; Jung, I.; Tutuc, E.S.; Banerjee, K.; Colombo, L.; Ruoff, R.S. Observation of single colloidal platinum nanocrystal growth trajectories. *Science,* **2009**, *324*(5932), 1309-1312.
[http://dx.doi.org/10.1126/science.1172104] [PMID: 19498166]

[21]   Berger, C.; Song, Z.; Li, T.; Li, X.; Ogbazghi, A.Y.; Feng, R.; Dai, Z.; Marchenkov, A.N.; Conrad, E.H.; First, P.N.; de Heer, W.A. Ultrathin epitaxial graphite: 2D electron gas properties and a route toward graphene based nanoelectronics. *J. Phys. Chem. B,* **2004**, *108*(52), 19912-19916.
[http://dx.doi.org/10.1021/jp040650f]

[22]   Cano-Márquez, A.G.; Rodríguez-Macías, F.J.; Campos-Delgado, J.; Espinosa-González, C.G.; Tristán-López, F.; Ramírez-González, D.; Cullen, D.A.; Smith, D.J.; Terrones, M.; Vega-Cantú, Y.I. Ex-MWNTs: graphene sheets and ribbons produced by lithium intercalation and exfoliation of carbon nanotubes. *Nano Lett.,* **2009**, *9*(4), 1527-1533.
[http://dx.doi.org/10.1021/nl803585s] [PMID: 19260705]

[23]   Brodie, B.C. On the atomic weight of graphite. *Philos. Trans. R. Soc,* **1859**, *149*, 249-259.
[http://dx.doi.org/10.1098/rstl.1859.0013]

[24]   Staudenmaier, L. Process for the preparation of graphitic acid. *Eur. J. Inorg. Chem.,* **1898**, *31*(2), 1481-1477.

[25]   Hummers, W.S.; Offeman, R.E. Preparation of graphitic oxide. *J. Am. Chem. Soc.,* **1958**, *80*(6), 1339-1339.
[http://dx.doi.org/10.1021/ja01539a017]

[26]   Cote, L.J.; Cruz-Silva, R.; Huang, J. Flash reduction and patterning of graphite oxide and its polymer composite. *J. Am. Chem. Soc.,* **2009**, *131*(31), 11027-11032.
[http://dx.doi.org/10.1021/ja902348k] [PMID: 19601624]

[27]   Zhou, Y.; Bao, Q.; Tang, L.; Zhong, Y.; Loh, K.P. Hydrothermal dehydration for the "Green" reduction of exfoliated graphene oxide to graphene and demonstration of tunable optical limiting properties. *Chem. Mater.,* **2009**, *21*(13), 2950-2956.
[http://dx.doi.org/10.1021/cm9006603]

[28]   Ai, K.; Liu, Y.; Lu, L.; Cheng, X.; Huo, L. A novel strategy for making soluble reduced graphene

oxide sheets cheaply by adopting an endogenous reducing agent. *J. Mater. Chem.,* **2011**, *21*(10), 3365-3370.
[http://dx.doi.org/10.1039/C0JM02865G]

[29]   Ramesha, G.K.; Sampath, S. Electrochemical reduction of oriented graphene oxide films: An insitu Raman spectroelectrochemical study. *J. Phys. Chem. C,* **2009**, *113*(19), 7985-7989.
[http://dx.doi.org/10.1021/jp811377n]

[30]   Xie, X. Zhou, Y.; Huang, K. Advances in Microwave-assisted production of reduced graphene oxide. *Front Chem.,* **2019**, *7*(355), 1-11.

[31]   Bose, S.; Drzal, L.T. Role of thickness and intercalated water in the facile reduction of graphene oxide employing camera flash. *Nanotechnology,* **2014**, *25*(7), 075702-075711.
[http://dx.doi.org/10.1088/0957-4484/25/7/075702] [PMID: 24451202]

[32]   Wang, Y.; Shi, Z.; Yin, J. Facile synthesis of soluble graphene *via* a green reduction of graphene oxide in tea solution and its biocomposites. *ACS Appl. Mater. Interfaces,* **2011**, *3*(4), 1127-1133.
[http://dx.doi.org/10.1021/am1012613] [PMID: 21438576]

[33]   Gurunathan, S.; Han, J.W.; Park, J.H.; Eppakayala, V.; Kim, J.H. Ginkgo biloba: a natural reducing agent for the synthesis of cytocompatible graphene. *Int. J. Nanomedicine,* **2014**, *9*, 363-377.
[http://dx.doi.org/10.2147/IJN.S53538] [PMID: 24453487]

[34]   Lee, G.; Kim, B.S. Biological reduction of graphene oxide using plant leaf extracts. *Biotechnol. Prog.,* **2014**, *30*(2), 463-469.
[http://dx.doi.org/10.1002/btpr.1862] [PMID: 24375994]

[35]   Khan, M.; Abdulhadi, H. Al-Marri, Khan, M.; Mohri, N.; Adil, S.F.; Al- Warthan, A.; Siddiqui, M.R.H.; Tahir, M.N. Pulicaria glutinosa plant extract: A green and eco-friendly reducing agent for the preparation of highly reduced graphene oxide. *RSC Advances,* **2014**, *4*, 24119-24125.
[http://dx.doi.org/10.1039/C4RA01296H]

[36]   Agharkar, M.; Kochrekar, S.; Hidouri, S.; Azeez, M.A. *Mater. Res. Bull.,* **2014**, *59*, 323-328.
[http://dx.doi.org/10.1016/j.materresbull.2014.07.051]

[37]   Suresh, D.; Nethravathi, P.C.; Udayabhanu, A.; Nagabhushana, H.; Sharma, S.C. Spinach assisted green reduction of graphene oxide and its antioxidant and dye absorption properties. *Ceram. Int.,* **2015**, *41*(3), 4810-4813.
[http://dx.doi.org/10.1016/j.ceramint.2014.12.036]

[38]   Bhattacharya, G.; Sas, S.; Wadhwa, S.; Mathur, A.; McLaughlin, J.; Sinha Roy, S. Aloe vera assisted facile green synthesis of reuced graphene oxide for electrochemical and dye removal applications. *RSC Advances,* **2017**, *7*, 26680-26688.
[http://dx.doi.org/10.1039/C7RA02828H]

[39]   Chegyang, L.; Zhuana, Z.; Jin, X.; Chen, Z. A facile and green preparation of reduced graphene oxide using Eucalyptus leaf extract. *Appl. Surf. Sci.,* **2017**, *422*, 469-474.
[http://dx.doi.org/10.1016/j.apsusc.2017.06.032]

[40]   Mungse, H.P.; Bhakuni, N.; Tripathi, D.; Sharma, O.P.; Sain, B.; Khatri, O.P. Fractional distribution of graphene oxide and its potential as an efficient and reusable solid catalyst for esterification reactions. *J. Phys. Org. Chem.,* **2014**, *27*, 944-951.
[http://dx.doi.org/10.1002/poc.3375]

[41]   Dhakshinamoorthy, A.; Alvaro, M.; Puche, M.; Fornes, V.; Garcia, H. Graphene oxide as catalyst for the acetalization of aldehydes at room temperature. *ChemCatChem,* **2012**, *4*, 2026-2030.
[http://dx.doi.org/10.1002/cctc.201200461]

[42]   Roy, B.; Sengupta, D.; Basu, B. Graphene oxide (GO) catalyzed chemoselective thioacetalization of aldehydes under solvent-free conditions. *Tet. Lett.,* **2014**, *55*(48), 6596-6600.
[http://dx.doi.org/10.1016/j.tetlet.2014.10.043]

[43]   Yu, H.; Wang, X.; Zhu, Y.; Zhuang, G.; Zhong, X.; Wang, J.G. Solvent-free catalytic dehydrative

esterification of benzaldehyde over graphene oxide. *Chem. Phys. Lett.,* **2013**, *583*, 146-150.
[http://dx.doi.org/10.1016/j.cplett.2013.08.011]

[44]   Nongbe, M.C.; Oger, N.; Ekou, T.; Ekou, L.B.; Yao, K.; Grognec, E.L.; Felpin, F.X. Graphene-catalyzed transacetalization under acid-free conditions. *Tet. Lett.,* **2016**, *57*(41), 4637-4639.
[http://dx.doi.org/10.1016/j.tetlet.2016.09.022]

[45]   Khalili, D. Graphene oxide: A promising carbocatalyst for the regioselective thiocynation of aromatic amines, phenols anisols, and enolizable ketones by hydrogen peroxide/KSCN in water. *New J. Chem.,* **2016**, *40*, 2549-2553.
[http://dx.doi.org/10.1039/C5NJ02314A]

[46]   Zhu, S.; Chem, C.; Xue, Y.; Wu, J.; Wang, J. Graphene oxide: An efficient acid catalyst for alcoholysis and esterification reactions. *ChemCatChem,* **2014**, *6*, 3080-3083.
[http://dx.doi.org/10.1002/cctc.201402574]

[47]   Bhattacharya, S.; Ghosh, P.; Basu, B. Graphene oxide (GO) catalyzed transamidation of aliphatic amides: An efficient metal-free procedure. *Tet. Lett.,* **2018**, *59*(10), 899-903.
[http://dx.doi.org/10.1016/j.tetlet.2018.01.060]

[48]   Basu, B.; Kundu, S.; Sengupta, D. Graphene oxide as a carbocatalyst: The first example of a one-pot sequential dehydration-hydrothiolation of secondary aryl alcohols. *RSC Advances,* **2013**, *3*, 22130-22134.
[http://dx.doi.org/10.1039/c3ra44712j]

[49]   Dhakshinamoorthy, A.; Alvaro, M.; Concepción, P.; Fornés, V.; Garcia, H. Graphene oxide as an acid catalyst for the room temperature ring opening of epoxides. *Chem. Commun. (Camb.),* **2012**, *48*(44), 5443-5445.
http://dx.doi.org/10.1039/c2cc31385e] [PMID: 22534622]

[50]   Cao, Y.; Luo, X.; Yu, H.; Peng, F.; Wang, H.; Ning, G. sp2- and sp3- hybridized carbon materials as catalysts for aerobic oxidation of cyclohexane. *Catal. Sci. Technol.,* **2013**, *3*(10), 2654-2660.
[http://dx.doi.org/10.1039/c3cy00256j]

[51]   Jia, H.P.; Dreyer, D.R.; Bielawski, C.W. C-H oxidation using graphite oxide. *Tetrahedron,* **2011**, *67*(24), 4431-4434.
[http://dx.doi.org/10.1016/j.tet.2011.02.065]

[52]   Dreyer, D.R.; Jia, H.P.; Todd, A.D.; Geng, J.; Bielawski, C.W. Graphite oxide: a selective and highly efficient oxidant of thiols and sulfides. *Org. Biomol. Chem.,* **2011**, *9*(21), 7292-7295.
[http://dx.doi.org/10.1039/c1ob06102j] [PMID: 21909587]

[53]   Aghayan, M.M.; Boukherroub, R.; Nemati, M.; Rahimifard, M. Graphite oxide mediated oxidative aromatization of 1,4-dihydropyridines into pyridine derivatives. *Tet. Lett.,* **2012**, *53*(19), 2473-2475.
[http://dx.doi.org/10.1016/j.tetlet.2012.03.026]

[54]   Aghayan, M.M.; Azar, E.K.; Boukherroub, R. Graphite oxide: An efficient reagent for oxidation of alcohols under sonication. *Tet. Lett.,* **2012**, *53*(37), 4962-4965.
[http://dx.doi.org/10.1016/j.tetlet.2012.07.016]

[55]   Yang, J.H.; Sun, G.; Gao, Y.; Zhao, H.; Tang, P.; Tan, J.; Lu, A.H.; Ma, D. Direct catalytic oxidation of benzene to phenol over metal-free graphene-based catalyst. *Energy Environ. Sci.,* **2013**, *6*(3), 793-798.
[http://dx.doi.org/10.1039/c3ee23623d]

[56]   Li, C.; Li, L.; Sun, L.; Pei, Z.; Xie, J.; Zhang, S. Transformation of hydroquinone to benzoquinone mediated by reduced graphene oxide in aqueous solution. *Carbon,* **2015**, *89*, 74-81.
[http://dx.doi.org/10.1016/j.carbon.2015.03.027]

[57]   Bai, L.S.; Gao, X.M.; Zhang, X.; Sun, F.F.; Ma, W. Reduced graphene oxide as a recyclable catalyst for dehydrogenation of hydrazo compounds. Tet. *Lett,* **2014**, *55*(33), 4545-4558.

[58]   Chu, X.; Zhu, Q.; Dai, W-L.; Fan, K. Excellent catalytic performance of graphite oxide in the selective

oxidation of glutaraldehyde by aqueous hydrogen peroxide. *RSC Advances,* **2012**, *2*, 7135-7139.
[http://dx.doi.org/10.1039/c2ra21068a]

[59]   Huang, H.; Huang, J.; Liu, Y-M.; He, H-Y.; Cao, Y.; Fan, K-N. Graphite oxide as an efficient and durable metal-free catalyst for aerobic oxidative coupling of amines to imines. *Green Chem.,* **2012**, *14*, 930-934.
[http://dx.doi.org/10.1039/c2gc16681j]

[60]   Gao, Y.; Ma, D.; Wang, C.; Guan, J.; Bao, X. Reduced graphene oxide as a catalyst for hydrogenation of nitrobenzene at room temperature. *Chem. Commun. (Camb.),* **2011**, *47*(8), 2432-2434.
[http://dx.doi.org/10.1039/C0CC04420B] [PMID: 21170437]

[61]   Pehrun, T.I.; Bychko, I.B.; Trypolsky, A.I.; Strizhak, P.E. Catalytic properties of graphene material in the hydrogenation of ethylene. *Theor. Exp. Chem.,* **2013**, *48*(6), 367-370.
[http://dx.doi.org/10.1007/s11237-013-9282-1]

[62]   Zhang, J.; Chloen, S.; Chen, F.; Xu, W.; Deng, G.J.; Gong, H. Dehydrogenation of nitrogen heterocycles using graphene oxide as versatile metal-free catalyst under air. *Adv. Synth. Catal.,* **2017**, *359*, 2358-2363.
[http://dx.doi.org/10.1002/adsc.201700178]

[63]   Mahata, S.; Sahu, A.; Shukla, P.; Rai, A.; Singh, M.; Rai, V.K. The novel and efficient reduction of graphene oxide using *Ocimum sanctum* L. leaf extract as an alternative renewable bio-resource. *New J. Chem.,* **2018**, *42*(24), 19945-19952.
[http://dx.doi.org/10.1039/C8NJ04086A]

[64]   Wu, S.; He, Q.; Zhou, C.; Qi, X.; Huang, X.; Yin, Z.; Yang, Y.; Zhang, H. Synthesis of Fe_3O_4 and Pt nanoparticles on reduced graphene oxide and their use as a recyclable catalyst. *Nanoscale,* **2012**, *4*(7), 2478-2483.
[http://dx.doi.org/10.1039/c2nr11992g] [PMID: 22388949]

[65]   Verma, S.; Mungse, H.P.; Kumar, N.; Choudhary, S.; Jain, S.L.; Sain, B.; Khatri, O.P. Graphene oxide: an efficient and reusable carbocatalyst for aza-Michael addition of amines to activated alkenes. *Chem. Commun. (Camb.),* **2011**, *47*(47), 12673-12675.
[http://dx.doi.org/10.1039/c1cc15230k] [PMID: 22039588]

[66]   Majumder, B.; Sarma, D.; Bhattacharya, T.; Sarma, T.K. Graphene oxide as metal-free catalyst in oxidative dehydrogenative C-N coupling leading to a α-ketoamides: Importance of dual catalytic activity. *ACS Sustain. Chem.& Eng.,* **2017**, *5*, 9286-9294.
[http://dx.doi.org/10.1021/acssuschemeng.7b02267]

[67]   Gao, Y.; Tang, P.; Zhou, H.; Zhang, W.; Yang, H.; Yan, N.; Hu, G.; Mei, D.; Wang, J.; Ma, D. Graphene oxide catalyzed C-H bond activation: The importance of oxygen functional groups for biaryl construction. *Angew. Chem. Int. Ed. Engl.,* **2016**, *55*(9), 3124-3128.
[http://dx.doi.org/10.1002/anie.201510081] [PMID: 26809892]

[68]   Wu, H.; Su, C.; Tandiana, R.; Liu, C.; Qiu, C.; Bao, Y.; Wu, J.; Xu, Y.; Lu, J.; Fan, D.; Loh, K.P. Graphene-oxide catalyzed direct CH-CH type cross coupling: The intrinsic catalytic activities of zigzag edges. *Angew. Chem. Int. Ed. Engl.,* **2018**, *57*(34), 10848-10853.
[http://dx.doi.org/10.1002/anie.201802548] [PMID: 29749675]

[69]   Scheuermann, G.M.; Rumi, L.; Steurer, P.; Bannwarth, W.; Mülhaupt, R. Palladium nanoparticles on graphite oxide and its functionalized graphene derivatives as highly active catalysts for the Suzuki-Miyaura coupling reaction. *J. Am. Chem. Soc.,* **2009**, *131*(23), 8262-8270.
[http://dx.doi.org/10.1021/ja901105a] [PMID: 19469566]

[70]   Rumi, L.; Scheuermann, G.M.; Mülhaupt, R.; Bannwarth, W. Palladium nanoparticles on graphite oxide as catalyst for Suzuki-Miyaura, Mizoroki-Heck, and Sonogashira reactions. *Helv. Chim. Acta,* **2011**, *94*, 966-976.
[http://dx.doi.org/10.1002/hlca.201000412]

[71]   Kumar, V.A.; Rama Rao, K. Recyclable graphite oxide catalyzed Friedel-crafts addition of indoles to

α, β-unsaturated ketones. Tet. *Lett,* **2011**, *52*(40), 5188-5191.

[72]   Hu, F.; Patel, M.; Luo, F.; Flach, C.; Mendelsohn, R.; Garfunkel, E.; He, H.; Szostak, M. Graphene-catalyzed direct Friedel-Crafts alkylation reactions: Mechanism, selectivity and synthetic utility. *J. Am. Chem. Soc.,* **2015**, *137*(45), 14473-14480.
[http://dx.doi.org/10.1021/jacs.5b09636] [PMID: 26496423]

[73]   Acocella, M.R.; Mauro, M.; Guerra, G. Regio- and enantioselective friedel-crafts reactions of indoles to epoxides catalyzed by graphene oxide: a green approach. *ChemSusChem,* **2014**, *7*(12), 3279-3283.
[http://dx.doi.org/10.1002/cssc.201402770] [PMID: 25328083]

[74]   Wu, T.; Wang, X.; Qui, H.; Gao, J.; Wang, W.; Liu, Y. Graphene oxide reduced and modified by soft nanoparticles and its catalysis of the Knoevenagel condensation. *J. Mater. Chem.,* **2012**, *22*(11), 4772-4779.
[http://dx.doi.org/10.1039/c2jm15311d]

[75]   Kim, Y.; Some, S.; Lee, H. Graphene oxide as a recyclable phase transfer catalyst. *Chem. Commun. (Camb.),* **2013**, *49*(50), 5702-5704.
[http://dx.doi.org/10.1039/c3cc42787k] [PMID: 23689290]

[76]   Qi, G.; Zhang, W.; Dai, Y. An efficient synthesis of 5-substituted 1H-tetrazoles catalyzed by graphene. *Res. Chem. Intermed.,* **2015**, *41*(2), 1149-1155.
[http://dx.doi.org/10.1007/s11164-013-1260-7]

[77]   Kumari, S.; Shekhar, A.; Mungse, H.P.; Khatri, O.P.; Pathak, D.D. Metal free one-pot synthesis of amides using graphene oxide an efficient catalyst. *RSC Advances,* **2014**, *4*(78), 41690-41695.
[http://dx.doi.org/10.1039/C4RA07589G]

[78]   Aghayan, M.M.; Tavana, M.M.; Boukherroub, R. Direct oxidative synthesis of nitrones from aldehydes and primary anilines using graphite oxide and oxazone. Tet. *Lett,* **2014**, *55*(40), 5471-5474.

[79]   Dhopte, K.B.; Raut, D.S.; Patwardhan, A.V.; Nemade, P.R. Graphene oxide as recyclable catalyst for one-pot synthesis of α-aminophosphonates. *Synth. Commun.,* **2015**, *45*(6), 778-788.
[http://dx.doi.org/10.1080/00397911.2014.989447]

[80]   Khalili, D. Graphene oxide: a reusable and metal-free carbocatalyst for the one-pot synthesis of 2-amino-3-cyanopyridines. Tet. *Lett,* **2016**, *57*(15), 1721-1723.
[http://dx.doi.org/10.1016/j.tetlet.2016.03.020]

[81]   Li, J.; Wang, M.; Zhang, Y.; Fan, Z.; Zhang, W.; Sun, F.; Ma, N. Deaomatizing naphthol Mannich bases towards spiro thiazolidinethiones catalyzed by recycled reduced graphene oxide with air as oxidant. *ACS Sustain. Chem.& Eng.,* **2016**, *4*, 3489-3495.

[82]   Girish, Y.R.; Pandit, S.; Pandit, S.; De, M. Graphene oxide as a carbocatalyst for a Diels-Alder reaction in an aqueous medium. *Chem. Asian J.,* **2017**, *12*(18), 2393-2398.
[http://dx.doi.org/10.1002/asia.201701072] [PMID: 28815919]

[83]   Bhattacharya, S.; Ghosh, P.; Basu, B. Graphene oxide: An efficient carbocatalyst for the benign synthesis of functionalized 1,4-benzothiazines. Tet. *Lett,* **2017**, *58*(10), 926-931.
[http://dx.doi.org/10.1016/j.tetlet.2017.01.068]

[84]   Kausar, N.; Roy, I.; Chattopadhyay, D.; Das, A.R. Synthesis of 2,3-dihydroquinolinones and quinazolin-4(3H)-ones catalyzed by graphene oxide nanosheets in an aqueous medium. "On water" synthesis accompanied by carbocatalysis and selective C-C bond cleavage. *RSC Advances,* **2016**, *6*(27), 22320-22330.
[http://dx.doi.org/10.1039/C6RA00388E]

[85]   Mahata, S.; Sahu, A.; Shukla, P.; Rai, A.; Singh, M.; Rai, V.K. Graphene oxide catalyzed C-N/-S/[3+2] cyclization cascade for green synthesis of Thiazolidinone in water. *Lett. Org. Chem.,* **2018**, *15*, 665-672.
[http://dx.doi.org/10.2174/1570178614666171002145250]

SUBJECT INDEX

A

Acetylation Reaction 184
Acid(s) 8, 16, 24, 25, 26, 27, 28, 33, 36, 61,
70, 90, 140, 183, 184, 190, 193, 199
  aldehydes and mercapto acetic 199
  aryl boronic 90
  BOC-amino 27
  boric 193
  carboxylic 26, 184
  citric 28
  functionalized graphene oxide 25
  glutaric 190
  levulinic 140
  mercapto acetic 199
  nitric 8
  organic 183
  phenylboronic 70
  propionic 184
  pyrenebutyric 16
  sulfanilic 28
  Tetrachloropalladinic 61
Acidic catalytic sites 99
Activity 36, 58, 59, 63, 67, 83, 87, 107, 134,
145, 165, 169
  catalytic reduction 134
Adhesive tape method 11
Adsorption 3, 45, 121, 122, 123, 124, 125
  capacity, high 3
  energies 45, 121, 122, 125
  mechanism 124
  process 123
Agent 66, 86
  reducing-cum-stabilizing 86
Alcoholysis Reaction 186
Alkaline 43, 45, 47
  electrolyte 47
  fuel cell 43, 45
Alkynes 87, 142, 143, 144, 190
  aliphatic 87
  hydrogenation of 142, 143, 144, 190
Aminopyridine moieties 35

Anti-corrosion ability 3
Applications 7, 78, 85, 116, 117, 163
  electrocatalyst sensor 116, 117
  electrochemical 85
  energy conversion 163
  environmental 78
  sensor 116
  spintronic 7
Aromatic 5, 32, 84, 189
  alcohols reactions 32
  compounds reactions 32
  halides, deficient 84
  hydrazo compounds 189
  hydrocarbons 5
Asymmetric aldol reactions 27
Atmospheres, chamber's reaction 10
Atomic 5, 84, 132, 134, 155, 157
  absorption spectroscopy 84
  force microscope (AFM) 5, 132, 134, 155,
157
Auger electron spectroscopy 108
Azide-alkyne cycloaddition (AAC) 84

B

Band gap 12, 32, 99, 161
  altered energy 32
  energy 161
Benzyl bromide derivatives 86
Bimetallic 62, 63, 64, 69
  catalysts 62, 63, 64, 69
  nanoparticles 64
Binding energy 43, 45, 83, 121, 122
Bulk 82, 116
  analysis 82
  graphite method 116

C

Calculations 69, 115, 116, 122, 125, 126
  computational 125
  theoretical 69, 115, 122, 126

www.ingramcontent.com/pod-product-compliance
Lightning Source LLC
Chambersburg PA
CBHW050839220326
41598CB00006B/397